W9-CIB-617

PENGUIN BOOKS

The Whispering Land

Gerald Durrell was born in Jamshedpur, India, in 1925. He returned to England in 1928 before settling on the island of Corfu with his family. In 1945 he joined the staff of Whipsnade Park as a student keeper, and in 1947 he led his first animal-collecting expedition to the Cameroons. He later undertook numerous further expeditions, visiting Paraguay, Argentina, Sierra Leone, Mexico, Mauritius, Assam and Madagascar. His first television programme, *Two in the Bush*, which documented his travels to New Zealand, Australia and Malaya, was made in 1962; he went on to make seventy programmes about his trips around the world. In 1959 he founded the Jersey Zoological Park, and in 1963 he founded the Jersey Wildlife Preservation Trust. He was awarded the OBE in 1982. Encouraged to write about his life's work by his novelist brother Lawrence, Durrell published his first book, *The Overloaded Ark*, in 1953. It soon became a bestseller and he went on to write thirty-six other titles, including *My Family and Other Animals*, *The Bafut Beagles*, *Encounters with Animals*, *The Drunken Forest*, *A Zoo in My Luggage*, *The Whispering Land*, *Menagerie Manor*, *The Amateur Naturalist* and *The Aye-Aye and I*. Gerald Durrell died in 1995.

JUN 2 4 2013

591.9827 Dur

Durrell, G.
The whispering land.

PRICE: $16.00 (3559/he   )

JUN 2 2 2013

The Whispering Land

GERALD DURRELL

WITH ILLUSTRATIONS BY
Ralph Thompson

PENGUIN BOOKS

PENGUIN BOOKS

Published by the Penguin Group
Penguin Books Ltd, 80 Strand, London WC2R ORL, England
Penguin Group (USA) Inc., 375 Hudson Street, New York, New York 10014, USA
Penguin Group (Canada), 90 Eglinton Avenue East, Suite 700, Toronto, Ontario, Canada M4P 2Y3
(a division of Pearson Penguin Canada Inc.)
Penguin Ireland, 25 St Stephen's Green, Dublin 2, Ireland
(a division of Penguin Books Ltd)
Penguin Group (Australia), 250 Camberwell Road,
Camberwell, Victoria 3124, Australia (a division of Pearson Australia Group Pty Ltd)
Penguin Books India Pvt Ltd, 11 Community Centre,
Panchsheel Park, New Delhi – 110 017, India
Penguin Group (NZ), 67 Apollo Drive, Rosedale, Auckland 0632, New Zealand
(a division of Pearson New Zealand Ltd)
Penguin Books (South Africa) (Pty) Ltd, Block D, Rosebank Office Park, 181 Jan Smuts Avenue,
Parktown North, Gauteng 2193, South Africa

Penguin Books Ltd, Registered Offices: 80 Strand, London WC2R ORL, England

www.penguin.com

First published in Great Britain by Rupert Hart-Davis 1961
First published in the United States of America by The Viking Press 1962
Published in Penguin Books 1964
Reissued in this edition 2012

001

Copyright © Gerald Durrell, 1961
All rights reserved

Typeset in 11/13pt Dante MT by Palimpsest Book Production Limited,
Falkirk, Stirlingshire
Printed in Great Britain by Clays Ltd, St Ives plc

Except in the United States of America, this book is sold subject
to the condition that it shall not, by way of trade or otherwise, be lent,
re-sold, hired out, or otherwise circulated without the publisher's
prior consent in any form of binding or cover other than that in
which it is published and without a similar condition including this
condition being imposed on the subsequent purchaser

ISBN: 978-0-241-95584-0

www.greenpenguin.co.uk

MIX
Paper from
responsible sources
FSC™ C018179

Penguin Books is committed to a sustainable
future for our business, our readers and our planet.
This book is made from Forest Stewardship
Council™ certified paper.

ALWAYS LEARNING   **PEARSON**

THIS IS FOR

BEBITA

who, by leaving Argentina,
has deprived me of my best reason for returning

In calling up images of the past, I find the plains of Patagonia frequently pass across my eyes; yet these plains are pronounced by all to be wretched and useless. They can be described only by negative characters; without habitation, without water, without trees, without mountains, they support merely a few dwarf plants. Why, then, and the case is not peculiar to myself, have these arid wastes taken so firm a hold on my memory?

CHARLES DARWIN
THE VOYAGE OF H.M.S. BEAGLE

Contents

A Word in Advance

Some time ago I wrote a book (called *A Zoo in My Luggage*) in which I explained that after travelling to different parts of the world for a number of years, collecting live animals for various zoos, I became bored.

I was not bored, I hasten to add, with the expeditions, still less with the animals I found. I was bored with having to part with these animals when I returned to England. The only answer to this was to start my own zoo, and how I set out to West Africa to gather the nucleus for this project, brought them home, and eventually founded my zoo on the island of Jersey, in the Channel Islands, I told about in *A Zoo in My Luggage*.

This, then, is really a sort of continuation of that book, for in this I describe how my wife and I, accompanied by my indefatigable secretary, Sophie, went to spend eight months in Argentina, in order to bring back a nice South American collection for the Jersey Zoo, and how, in spite of many setbacks, this was what we did. If any praise for the collection is due it must go to Sophie, for, although she does not figure largely in these pages, she bore perhaps the greatest burden of the trip. Uncomplainingly she stayed in Buenos Aires and looked after the incessant flow of animals which I kept reappearing with from various places, and looked after them, moreover, in a way that would have done credit to a veteran collector. For this, and for many other reasons, I am deeply in her debt.

PART ONE

The Customs of the Country

Buenos Aires, decked out for spring, was looking her best. The tall and elegant buildings seemed to gleam like icebergs in the sun, and the broad avenues were lined with jacaranda trees covered with a mist of mauvy blue flowers, or *palo borracho*, with their strange bottle-shaped trunks and their spindly branches starred with yellow and white flowers. The spring-like atmosphere seemed to have infected the pedestrians, who fled across the road through the traffic with even less caution than usual, while the drivers of the trams, buses and cars vied with each other in the time-honoured Buenos Aires game of seeing how close they could get to each other at the maximum speed without actually crashing.

Not having a suicidal streak in me, I had refused to drive in the city, and so we swept on our death-defying way in the Land-Rover with Josefina at the wheel. Short, with curly auburn hair and big brown eyes, Josefina had a smile like a searchlight that could paralyse even the most unsusceptible male at twenty paces. By my side sat Mercedes, tall, slim, blonde and blue-eyed; she habitually wore an expression as though butter would not melt in her mouth, and this successfully concealed an iron will and grim, bulldog-like tenacity of purpose. These two girls were part of my private army of feminine pulchritude that I used in dealing with officialdom in the Argentine. At that precise moment we were heading towards the massive building that looked like a cross between the Parthenon and the Reichstag in whose massive interior lurked the most formidable enemy of sanity and liberty in Argentina: the Aduana, or Customs. On my arrival, some three weeks earlier, they had let all my highly dutiable articles of equipment, such as cameras, film, the Land-Rover and so on, into the country without a murmur; but, for some reason known only to the Almighty and the scintillating brains in the Aduana, they had confiscated all my nets, traps, cage-fronts and other worthless but necessary items of collecting equipment. So, for the past three weeks Mercedes, Josefina and I had spent every day in the bowels of the massive Customs House, being passed from office to office with a sort of clockwork-like regularity which was so monotonous and so frustrating that you really began to wonder if your brain would last out the course. Mercedes regarded me anxiously as Josefina wove in and out of fleeing pedestrians in a way that made my stomach turn over.

'How are you feeling today, Gerry?' she asked.

'Wonderful, simply wonderful,' I said bitterly; 'there's nothing I like better than to get up on a lovely morning like this and to feel that I have the whole sunlit day lying ahead in which to get on more intimate terms with the Customs.'

'Now, please don't talk like that,' she said; 'you promised me you wouldn't lose your temper again; it doesn't do any good.'

'It may not do any good, but it relieves my feelings. I swear to you that if we are kept waiting half an hour outside an office to be told by its inmate at the end of it that it's not *his* department, and to go along to Room Seven Hundred and Four, I shall not be responsible for my actions.'

'But today we are going to see Señor Garcia,' said Mercedes, with the air of one promising a sweet to a child.

I snorted. 'To the best of my knowledge we have seen at least fourteen Señor Garcias in that building in the past three weeks. The Garcia tribe treat the Customs as though it's an old family firm. I should imagine that all the baby Garcias are born with a tiny rubber-stamp in their hands,' I said, warming to my work. 'As christening presents they receive faded portraits of San Martin, so that when they grow up they can hang them in their offices.'

'Oh dear, I think you'd better sit in the car,' said Mercedes.

'What, and deprive me of the pleasure of continuing my genealogical investigation of the Garcia family?'

'Well, promise that you won't say anything,' she said, turning her kingfisher-blue eyes on me pleadingly. 'Please, Gerry, not a word.'

'But I never do say anything,' I protested, 'if I really voiced my thoughts the whole building would go up in flames.'

'What about the other day when you said that under the dictatorship you got your things in and out of the country without trouble, whereas now we were a democracy you were being treated like a smuggler?'

'Well, it's perfectly true. Surely one is allowed to voice one's thoughts, even in a democracy? For the last three weeks we have done nothing but struggle with these moronic individuals in the Customs, none of whom appears to be able to say anything except advise you to go and see Señor Garcia down the hall. I've wasted three weeks of valuable time when I could have been filming and collecting animals.'

'De hand . . . de hand . . .' Josefina said suddenly and loudly. I stuck my arm out of the window, and the speeding line of traffic behind us screeched to a shuddering halt as Josefina swung the Land-Rover into the side turning. The shouts of rage with cries of '*¡animál!*' faded behind us.

'Josefina, I do wish you would give us all a little more warning when you're going to turn,' I said. Josefina turned her glittering smile on to me.

'Why?' she inquired simply.

'Well, it helps you know. It gives us a chance to prepare to meet our Maker.'

'I 'ave never crash you yet, no?' she asked.

'No, but I feel it's only a matter of time.'

We swept majestically across an intersection at forty miles an hour, and a taxi coming from the opposite direction had to apply all its brakes to avoid hitting us amidships.

'Blurry Bas-tard,' said Josefina tranquilly.

'Josefina! You must not use phrases like that,' I remonstrated.

'Why not?' asked Josefina innocently. 'You do.'

'That is not the point,' I said severely.

'But it is nice to say, no?' she said with satisfaction. 'And I 'ave learn more; I know Blurry Bastard and . . .'

'All right, all right,' I said hastily. 'I believe you. But for Heaven's sake don't use them in front of your mother, otherwise she'll stop you driving for me.'

There were, I reflected, certain drawbacks to having beautiful young women to help you in your work. True, they could charm the birds out of the trees, but I found that they also had tenacious memories when it came to the shorter, crisper Anglo Saxon expletives which I was occasionally driven to using in moments of stress.

'De hand . . . de hand,' said Josefina again, and we swept across the road, leaving a tangle of infuriated traffic behind us, and drew up outside the massive and gloomy façade of the Aduana.

Three hours later we emerged, our brains numb, our feet aching, and threw ourselves into the Land-Rover.

'Where we go to now?' inquired Josefina listlessly.

'A bar,' I said, 'any bar where I can have a brandy and a couple of aspirins.'

'O.K.,' said Josefina, letting in the clutch.

'I think tomorrow we will have success,' said Mercedes, in an effort to revive our flagging spirits.

'Listen,' I said with some asperity, 'Señor Garcia, God bless his blue chin and eau-de-cologne-encrusted brow, was about as much use as a beetle in a bottle. And you know it.'

'No, no, Gerry. He has promised tomorrow to take me to see one of the high-up men in the Aduana.'

'What's *his* name . . . Garcia?'

'No, a Señor Dante.'

'How singularly appropriate. Only a man with a name like Dante would be able to survive in that Inferno of Garcias.'

'And you nearly spoilt everything,' said Mercedes reproachfully, 'asking him if that was a picture of his father. You knew it was San Martin.'

'Yes, I know, but I felt if I didn't say something silly my brain would snap like a pair of ancient elastic-sided boots.'

Josefina drew up outside a bar, and we assembled at a table on the edge of the pavement and sipped our drinks in depressed silence. Presently I managed to shake my mind free of the numbing effect that the Aduana always had on it, and turn my attention to other problems.

'Lend me fifty cents, will you?' I asked Mercedes. 'I want to phone up Marie.'

'Why?' inquired Mercedes.

'If you must know she's promised to find me a place to keep the tapir. The hotel won't let me keep it on the roof.'

'What is a tapir?' asked Josefina interestedly.

'It's a sort of animal, about as big as a pony, with a long nose. It looks like a small elephant gone wrong.'

'I am not surprised that the hotel won't let you keep it on the roof,' said Mercedes.

'But this one's only a baby . . . about the size of a pig.'

'Well, here's your fifty cents.'

I found the phone, mastered the intricacies of the Argentine telephone system and dialled Marie's number.

'Marie? Gerry here. What luck about the tapir?'

'Well, my friends are away so you can't take him there. But Mama says why not bring him here and keep him in the garden.'

'Are you sure that's all right?'

'Well, it was Mama's idea.'

'Yes, but are you sure she knows what a tapir is?'

'Yes, I told her it was a little animal with fur.'

'Not exactly a zoological description. What's she going to say when I turn up with something that's nearly bald and the size of a pig?'

'Once it's here, it's here,' said Marie logically.

I sighed.

'All right. I'll bring it round this evening. O.K.?'

'O.K., and don't forget some food for it.'

I went back to where Josefina and Mercedes were waiting with an air of well-bred curiosity.

'Well, what did she say?' inquired Mercedes at length.

'We put Operation Tapir into force at four o'clock this afternoon.'

'Where do we take it?'

'To Marie's house. Her mother's offered to keep it in the garden.'

'Good God, no!' said Mercedes with considerable dramatic effect.

'Well, why not?' I asked.

'But you cannot take it there, Gerry. The garden is only a small one. Besides, Mrs Rodrigues is very fond of her flowers.'

'What's that got to do with the tapir? He'll be on a leash. Anyway, he's got to go somewhere, and that's the only offer of accommodation I've had so far.'

'All right, take him there,' said Mercedes with the ill-concealed air of satisfaction of one who knows she is right, 'but don't say I didn't warn you.'

'All right, all right. Let's go and have some lunch now, because I've got to pick up Jacquie at two o'clock to go and see the shipping people about our return passages. After that we can go and pick up Claudius.'

'Who's Claudius?' asked Mercedes, puzzled.

'The tapir. I've christened him that because with that Roman snout of his he looks like one of the ancient Emperors.'

'Claudius!' said Josefina, giggling. 'Dat is blurry funny.'

So, at four o'clock that afternoon we collected the somewhat reluctant tapir and drove round to Marie's house, purchasing en route a long dog-leash and a collar big enough for a Great Dane. The garden was, as Mercedes had said, very small. It measured some fifty feet by fifty, a sort of hollow square surrounded on three sides by the black walls of the neighbouring houses, and on the fourth side was a tiny verandah and French windows, leading into the Rodrigues establishment. It was, by virtue of the height of the building surrounding it, a damp and rather gloomy little garden, but Mrs Rodrigues had done wonders to improve it by planting those flowers and shrubs which flourish best in such ill-lit situations. We had to carry Claudius, kicking violently, through the house, out of the French windows, where we attached his leash to the bottom of the steps. He wiffled his Roman snout appreciatively at the scents of damp earth and flowers that were wafted to him, and heaved a deep sigh of content. I placed a bowl of water by his side, a huge stack of chopped vegetables and fruit, and left him. Marie promised that she would phone me at the hotel the first thing the following morning and let me know how Claudius had settled down. This she dutifully did.

'Gerry? Good morning.'

'Good morning. How's Claudius?'

'Well, I think you had better come round,' she said with an air of someone trying to break bad news tactfully.

'Why, what's the matter? He's not ill, is he?' I asked, alarmed.

'Oh, no. Not ill,' said Marie sepulchrally. 'But last night he broke his leash, and by the time we discovered him, he had eaten half Mama's begonias. I've got him locked in the coal cellar, and Mama's upstairs having a headache. I think you had better come round and bring a new leash.'

Cursing animals in general and tapirs in particular, I leapt into a taxi and fled round to Marie's, pausing on the way to buy four-teen pots of the finest begonias I could procure. I found Claudius, covered with coal-dust, meditatively chewing a leaf. I repri-manded him, put on his new and stronger leash (strong enough, one would have thought, to hold a dinosaur), wrote a note of apology to Mrs Rodrigues, and left, Marie having promised to get in touch immediately should anything further transpire. The next morning she rang me again.

'Gerry? Good morning.'

'Good morning. Everything all right?'

'No,' said Marie gloomily, 'he's done it again. Mama has no begonias left now, and the rest of the garden looks as if a bull-dozer's been at work. I think he will have to have a chain you know.'

'Dear God,' I groaned, 'what with the Aduana and this bloody tapir, it's enough to drive one to drink. All right, I'll come round and bring a chain.'

Once more I arrived at the Rodrigues establishment, carrying a chain that could have been used to anchor the *Queen Mary*, and bearing another herbaceous border in pots. Claudius was enchanted with the chain. He found it tasted very nice if sucked loudly, and better still, it made a loud and tuneful rattling if he jerked his head up and down, a noise that suggested there was a small iron-foundry at work in the Rodrigues garden. I left hur-riedly before Mrs Rodrigues came down to ascertain the cause of the noise. Marie phoned me the following morning.

'Gerry? Good morning.'

'Good morning,' I said, with a strong premonition that it was going to turn out to be anything but a good morning.

'I'm afraid Mama says you will have to move Claudius,' said Marie.

'What's he done *now*?' I asked in exasperation.

'Well,' said Marie, with the faintest tremor of mirth in her voice, 'Mama gave a dinner party last night. Just as we had all sat down there was a terrible noise in the garden. Claudius had managed to get his chain loose from the railings, I don't know how. Anyway, before we could do anything sensible he burst in through the French windows, dragging his chain behind him.'

'Good God!' I said, startled.

'Yes,' said Marie, starting to giggle helplessly, 'it was so funny. All the guests leaping about, quite terrified, while Claudius ran round and round the table, clanking his chain like a spectre. Then he got frightened at all the noise and did a . . . you know . . . a *decoration* on the floor.'

'Dear Heaven,' I groaned, for I knew what Claudius could do in the way of 'decoration' when he put his mind to it.

'So Mama's dinner was ruined, and she says she is very sorry, but could you move him. She feels that he is not happy in the garden, and that anyway, he's not a very *simpatico* animal.'

'Your mother is, I presume, upstairs having a headache?'

'I think it's a bit more than a headache,' said Marie judiciously.

'O.K.' I sighed, 'leave it to me. I'll think of something.'

This, however, appeared to be the last of a series of bedevilments we had suffered, for suddenly everything seemed to go right. The Customs released my equipment, and, more important still, I suddenly found not only a home for Claudius, but the rest of the animals as well: a small house on the outskirts of Buenos Aires had been lent to use to keep our collection in as a temporary measure.

So, with our problems solved, at least for the moment, we got out the maps and planned our route to the south, to the Patagonian coastline where the fur seals and elephant seals gambolled in the icy waters.

At first sight everything seemed to be quite straightforward. Marie had managed to obtain leave from her job, and was to come with us to act as interpreter. Our route was planned with the minute detail that only people who have never been to an area indulge in. The equipment was checked and double-checked, and carefully packed. After all the weeks of frustration and boredom in Buenos Aires we began to feel that at last we were on our way. Then, at our last council of war (in the little café on the corner), Marie produced an argument that she had obviously been brooding upon for some considerable time.

'I think it would be a good idea if we take someone who knows the roads, Gerry,' she said, engulfing what appeared to be a large loaf of bread stuffed with an exceptionally giant ox's tongue, a concoction that passed for a sandwich in Argentina.

'Whatever for?' I asked. 'We've got maps, haven't we?'

'Yes, but you have never driven on those Patagonian roads, and they are quite different from anywhere else in the world, you know.'

'How, different?' I inquired.

'Worse,' said Marie, who did not believe in wasting words.

'I'm inclined to agree,' said Jacquie. 'We've heard the most awful reports of those roads from everyone.'

'Darling, you know as well as I do that you *always* hear those sorts of reports about roads, or mosquitoes, or savage tribes, wherever you go in the world, and they are generally a lot of nonsense.'

'Anyway, I think Marie's suggestion is a good one. If we could get someone who knows the roads to drive us down, then you know what to expect on the way back.'

'But there *is* no one,' I said irritably, 'Rafael is in college, Carlos is up in the North, Brian is studying . . .'

'There is Dicky,' said Marie.

I stared at her.

'Who is Dicky?' I asked at length.

'A friend of mine,' she said carelessly, 'he is a very good driver, he knows Patagonia, and he is a very nice person. He is quite used to going on hunting trips, so he does not mind suffering.'

'By "suffering" do you mean roughing it, or are you insinuating that our company might be offensive to his delicate nature?'

'Oh, stop being facetious,' said Jacquie. 'Would this chap come with us, Marie?'

'Oh, yes,' she said. 'He said he would like it very much.'

'Good,' said Jacquie, 'when can he come and see us?'

'Well, I told him to meet us here in about ten minutes' time,' said Marie. 'I thought Gerry would want to see him in case he did not like him.'

I gazed at them all speechlessly.

'I think that's a very good idea, don't you?' asked Jacquie.

'Are you asking my opinion?' I inquired. 'I thought you had settled it all between you.'

'I am sure you will like Dicky . . .' began Marie, and at that moment Dicky arrived.

At first glance I decided that I did not like Dicky at all. He did not look to me the sort of person who had ever suffered, or, indeed, was capable of suffering. He was exquisitely dressed, too exquisitely dressed. He had a round, plump face, with boot-button eyes, a rather frail-looking moustache like a brown moth decorated his upper lip, and his dark hair was plastered down to his head with such care that it looked as if it had been painted on to his scalp.

'This is Dicky de Sola,' said Marie, in some trepidation.

Dicky smiled at me, a smile that transformed his whole face.

'Marie have told you?' he said, dusting his chair fastidiously with his handkerchief before sitting down at the table. 'I am delight to come with you if you are happy. I am delight to go to Patagonia, whom I love.'

I began to warm to him.

'If I am no useful, I will not come, but I can advise if you will

13

allow, for I know the roads. You have a map? Ah, good, now let me explanation to you.'

Together we pored over the map, and within half an hour Dicky had won me over completely. Not only did he have an intimate knowledge of the country we were to pass through, but his own brand of English, his charm and infectious humour had decided me.

'Well,' I said, as we folded the maps away, 'if you can really spare the time, we'd like you to come very much.'

'Overwhelmingly,' said Dicky, holding out his hand.

And on this rather cryptic utterance the bargain was sealed.

I.

The Whispering Land

The plains of Patagonia are boundless, for they are scarcely pass-
able, and hence unknown; they bear the stamp of having lasted,
as they are now, for ages, and there appears no limit to their
duration through future time.

CHARLES DARWIN: THE VOYAGE OF H.M.S. BEAGLE

We set off for the south in the pearly grey dawn light of what
promised to be a perfect day. The streets were empty and echoing, and the dew-drenched parks and squares had their edges
frothed with great piles of fallen blooms from the *palo borracho*
and jacaranda trees, heaps of glittering flowers in blue, yellow
and pink.

On the outskirts of the city we rounded a corner and came
upon the first sign of life we had seen since we had started, a
covey of dustmen indulging in their early morning ballet. This
was such an extraordinary sight that we drove slowly behind

them for some way in order to watch. The great dustcart rumbled down the centre of the road at a steady five miles an hour, and standing in the back, up to his knees in rubbish, stood the emptier. Four other men loped alongside the cart like wolves, darting off suddenly into dark doorways to reappear with dustbins full of trash balanced on their shoulders. They would run up alongside the cart and throw the dustbin effortlessly into the air, and the man on the cart would catch it and empty it and throw it back, all in one fluid movement. The timing of this was superb, for as the empty dustbin was hurtling downwards a full one would be sailing up. They would pass in mid air, and the full bin would be caught and emptied. Sometimes there would be four dustbins in the air at once. The whole action was performed in silence and with incredible speed.

Soon we left the edge of the city, just stirring to wakefulness, and sped out into the open countryside, golden in the rising sun. The early morning air was chilly, and Dicky had dressed for the occasion. He was wearing a long tweed overcoat and white gloves, and his dark, bland eyes and neat, butterfly-shaped moustache peered out from under a ridiculous deer-stalker hat which he wore, he explained to me, in order to 'keep the ears heated'. Sophie and Marie crouched in strange prenatal postures in the back of the Land-Rover, on top of our mountainous pile of equipment, most of which, they insisted, had been packed in boxes with knife-like edges. Jacquie and I sat next to Dicky in the front seat, a map spread out across our laps, our heads nodding, as we endeavoured to work out our route. Some of the places we had to pass through were delightful: Chascomus, Dolores, Necochea, Tres Arroyos, and similar delicious names that slid enticingly off the tongue. At one point we passed through two villages, within a few miles of each other, one called 'The Dead Christian' and the other 'The Rich Indian'. Marie's explanation of this strange nomenclature was that the Indian was rich because he killed the Christian, and had stolen all his money, but attractive though this story was, I felt it could not be the right one.

For two days we sped through the typical landscape of the Pampa, flat golden grassland in which the cattle grazed knee-deep; occasional clumps of eucalyptus trees, with their bleached and peeling trunks like leprous limbs; small, neat *estancias*, gleaming white in the shade of huge, carunculated *ombú* trees, that stood massively and grimly on their enormous squat trunks. In places the neat fences that lined the road were almost obliterated under a thick cloak of convolvulus, hung with electric-blue flowers the size of saucers, and every third or fourth fence-post would have balanced upon it the strange, football-like nest of an oven-bird. It was a lush, prosperous and well-fed-looking landscape that only just escaped being monotonous. Eventually, in the evening of the third day, we lost our way, and so we pulled in to the side of the road and argued over the map. Our destination was a town called Carmen de Patagones, on the north bank of the Rio Negro. I particularly wanted to spend the night here, because it was a town that Darwin had stayed in for some time during the voyage of the *Beagle*, and I was interested to see how it had changed in the last hundred years. So, in spite of near-mutiny on the part of the rest of the expedition, who wanted to stop at the first suitable place we came to, we drove on. As it turned out it was all we could have done anyway, for we did not pass a single habitation until we saw gleaming ahead of us a tiny cluster of feeble lights. Within ten minutes we were driving cautiously through the cobbled streets of Carmen de Patagones, lit by pale, trembling street-lights. It was two o'clock in the morning, and every house was blank-faced and tightly shuttered. Our chances of finding anyone who could direct us to a hostelry were remote, and we certainly needed direction, for each house looked exactly like the ones on each side of it, and there was no indication as to whether it was a hotel or a private habitation. We stopped in the main square of the town and were arguing tiredly and irritably over this problem when suddenly, under one of the street-lights, appeared an angel of mercy, in the shape of a tall, slim policeman clad in an immaculate uniform, his belt and boots gleaming. He saluted smartly, bowed to the female members of the party, and with old-world courtesy directed us up some side-roads to where he said we should find a hotel. We came to a great gloomy house, heavily shuttered, with a massive front door that would have done justice to a cathedral. We beat a sharp tattoo on its weather-beaten surface and awaited results patiently. Ten minutes later there was still no response from the inhabitants, and so Dicky, in desperation, launched an assault on the door that would, if it had succeeded, have awakened the dead. But as he lashed out at the door it swung mysteriously open under his assault, and displayed a long, dimly lit passageway, with doors along each side, and a marble staircase leading to the upper floors. Dead tired and extremely hungry we were in no mood to consider other people's property, so we marched into the echoing hall like an invading army. We stood there and shouted '¡Holà!' until the hotel rang with our shouts, but there was no response.

'I think, Gerry, that sometime they are all deceased,' said Dicky gravely.

'Well, if they are I suggest we spread out and find ourselves some beds,' I said.

So we climbed the marble staircase and found ourselves three bedrooms, with beds made up, by the simple expedient of opening every door in sight. Eventually, having found a place to sleep, Dicky and I went downstairs to see if the hotel boasted of any sanitary arrangements. The first door we threw open in our search led us into a dim bedroom in which was an enormous double-bed hung with an old-fashioned canopy. Before we could back out of the room a huge figure surged out from under the bedclothes like a surfacing whale, and waddled towards us. It turned out to be a colossal woman, clad in a flowing flannel nightie, who must have weighed somewhere in the neighbourhood of fifteen stone. She

came out, blinking into the hallway, pulling on a flowing kimono of bright green covered with huge pink roses, so the effect was rather as if one of the more exotic floral displays of the Chelsea Flower Show had suddenly taken on a life of its own. Over her ample bosoms spread two long streamers of grey hair which she flicked deftly over her shoulder as she did up her kimono, smiling at us with sleepy goodwill.

'*Buenas noches,*' she said politely.

'*Buenas noches, señora,*' we replied, not to be outdone in good manners at that hour of the morning.

'*¿Hablo con la patrona?*' inquired Dicky.

'*Si, si, señor,*' she said, smiling broadly, '*¿que quieres?*'

Dicky apologized for our late arrival, but *la patrona* waved away our apologies. Was it possible, Dicky asked, for us to have some sandwiches and coffee? Why not? inquired *la patrona*. Further, said Dicky, we were in urgent need of a lavatory, and could she be so kind as to direct us to it. With great good humour she led us to a small tiled room, showed us how to pull the plug, and stood there chatting amiably while Dicky and I relieved the pangs of nature. Then she puffed and undulated her way down to the kitchen and cut us a huge pile of sandwiches and made a steaming mug of coffee. Having assured herself that there was nothing further she could do for our comfort, she waddled off to bed.

The next morning, having breakfasted, we did a rapid tour of the town. As far as I could see, apart from the introduction of electricity, it had changed very little since Darwin's day, and so we left and sped down a hill and across the wide iron bridge that spanned the rusty red waters of the Rio Negro. We rattled across the bridge from the Province of Buenos Aires to the Province of Chubut, and by that simple action of crossing a river we entered a different world.

Gone were the lush green plains of the Pampa, and in their place was an arid waste stretching away as far as the eye could see on each side of the dusty road, a uniform pelt of grey-green scrub composed of plants about three feet high, each armed with a formidable array of thorns and spikes. Nothing appeared to live in this dry scrub, for when we stopped there was no bird or insect song, only the whispering of the wind through the thorn scrub in this monochromatic Martian landscape, and the only moving thing apart from ourselves was the giant plume of dust we trailed behind the vehicle. This was terribly tiring country to drive in. The road, deeply rutted and potholed, unrolled straight ahead to the horizon, and after a few hours this monotony of scene numbed one's brain, and one would suddenly drop off to sleep, to be awoken by the vicious scrunch of the wheels as the Land-Rover swerved off into the brittle scrub.

The evening before we were due to reach Deseado this happened on a stretch of road which, unfortunately, had recently been rained upon, so that the surface had turned into something resembling high-grade glue. Dicky, who had been driving for a long time, suddenly nodded off behind the wheel, and before anyone could do anything sensible, both Land-Rover and trailer had skidded violently into the churned-up mud at the side of the road, and settled there snugly, wheels spinning like mad. Reluctantly we got out into the bitter chill of the evening wind, and in the dim sunset light set to work to unhitch the trailer and then push it and the Land-Rover separately out of the mud. Then, our feet and hands frozen, the five of us crouched in the shelter of the Land-Rover and watched the sunset, passing from hand to hand a bottle of Scotch which I had been keeping for just such an emergency.

On every side of us the scrubland stretched away, dark and flat, so that you got the impression of being in the centre of a gigantic plate. The sky had become suffused with green as the sun sank, and then, unexpectedly, turned to a very pale powder-blue. A tattered mass of clouds on the western horizon suddenly turned black, edged delicately with flame-red, and resembled a great armada of Spanish galleons waging a fierce sea-battle across the sky, drifting towards each other, turned into black silhouette by the fierce glare from their cannons. As the sun sank lower and lower the black of the clouds became shot and mottled with grey, and the sky behind them became striped with green, blue and pale red. Suddenly our fleet of galleons disappeared, and in its place was a perfect archipelago of islands strung out across the sky in what appeared to be a placid, sunset-coloured sea. The illusion was perfect: you could pick out the tiny, white rimmed coves in the rocky, indented shoreline, the occasional long, white beach; the dangerous shoal of rocks formed by a wisp of cloud at the entrance to a safe anchorage; the curiously-shaped mountains inland covered with a tattered pelt of evening-dark forest. We sat there, the whisky warming our bodies, watching enraptured the geography of this archipelago unfold. We each of us chose an island which appealed to us, on which we would like to spend a holiday, and stipulated what the hotel on each of our islands would have to provide in the way of civilized amenities.

'A very, *very* big bath, and very deep,' said Marie.

'No, a nice hot shower and a comfortable chair,' said Sophie.

'Just a bed,' said Jacquie, 'a large feather bed.'

'A bar that serves real ice with its drinks,' I said dreamily.

Dicky was silent for a moment. Then he glanced down at his feet, thickly encrusted with rapidly drying mud.

'I must have a man to clean my feets,' he said firmly.

'Well, I doubt whether we'll get any of that at Deseado,' I said gloomily, 'but we'd better press on.'

When we drove into Deseado at ten o'clock the next morning, it became immediately obvious that we could not expect any such luxuries as feather beds, ice in the drinks, or even a man to clean our feets. It was the most extraordinarily dead-looking town I had ever been in. It resembled the set for a rather bad Hollywood cowboy film, and gave the impression that its inhabitants (two thousand, according to the guide-book) had suddenly packed up and left it alone to face the biting winds and scorching sun. The empty, rutted streets between the blank-faced houses were occasionally stirred by the wind, which produced half-hearted dust-devils, that swirled up for a moment and then collapsed tiredly to the ground. As we drove slowly into what we imagined to be the centre of the town we saw only a dog, trotting briskly about his affairs, and a child crouched in the middle of a road, absorbed in some mysterious game of childhood. Then, swinging the Land-Rover round a corner, we were startled to see a man on horseback, clopping slowly along the road with the subdued air of one who is the sole survivor of a catastrophe. He pulled up and greeted us politely, but without interest, when we stopped, and directed us to the only two hotels in the place. As these turned out to be opposite each other and both equally unprepossessing from the outside, we chose one by tossing a coin and made our way inside.

In the bar we found the proprietor, who, with the air of one who had just suffered a terrible bereavement, reluctantly admitted that he had accommodation, and led us through dim passages to three small, grubby rooms. Dicky, his deer-stalker on the back of his head, stood in the centre of his room, pulling off his white gloves, surveying the sagging bed and its grey linen with a cat-like fastidiousness.

'You know what, Gerry?' he said with conviction. 'This is the stinkiest hotel I ever dream.'

'I hope you never dream of a stinkier one,' I assured him.

Presently we all repaired to the bar to have a drink and await the arrival of one Captain Giri, whom I had an introduction to, a man who knew all about the penguin colonies of Puerto Deseado. We sat round a small table, sipping our drinks and watching the other inhabitants of the bar with interest. For the most part they seemed to consist of very old men, with long, sweeping moustaches, whose brown faces were seamed and stitched by the wind. They sat in small groups, crouched over their tiny tumblers of cognac or wine with a dead air, as though they were hibernating there in this dingy bar, staring hopelessly into the bottoms of their glasses, wondering when the wind would die down, and knowing it would not. Dicky, delicately smoking a cigarette, surveyed the smoke-blackened walls, the rows of dusty bottles, and the floor with its twenty-year-old layer of dirt well trodden into its surface.

'What a bar, eh?' he said to me.

'Not very convivial, is it?'

'It is so old . . . it has an air of old,' he said staring about him. 'You know, Gerry, I bet it is so old that even the flies have beards.'

Then the door opened suddenly, a blast of cold air rushed into the bar, the old men looked up in a flat-eyed, reptilian manner, and through the door strode Captain Giri. He was a tall, well-built man with blond hair, a handsome, rather aesthetic face and the most vivid and candid blue eyes I had ever seen. Having introduced himself he sat down at our table and looked round at us with such friendliness and good humour in his child-like eyes that the dead atmosphere of the bar dropped away, and we suddenly found ourselves becoming alive and enthusiastic. We had a drink, and then Captain Giri produced a large roll of charts and spread them on the table, while we pored over them.

'Penguins,' said the Captain meditatively, running his fore-finger over the chart. 'Now, down here is the best colony . . . by far the best and biggest, but I think that that is too far for you, is it not?'

'Well, it is a bit,' I admitted. 'We didn't want to go that far south if we could avoid it. It's a question of time, really. I had hoped that there would be a reasonable colony within fairly easy reach of Deseado.'

'There is, there is,' said the Captain, shuffling the charts like a conjuror and producing another one from the pile. 'Now, here, you see, at this spot . . . it's about four hours' drive from Deseado . . . all along this bay here.'

'That's wonderful,' I said enthusiastically, 'just the right distance.'

'There is only one thing that worries me,' said the Captain, turning troubled blue eyes on to me. 'Are there enough birds there for what you want . . . for your photography?'

'Well,' I said doubtfully, 'I want a fair number. How many are there in this colony?'

'At a rough estimate I should say a million,' said Captain Giri. 'Will that be enough?'

I gaped at him. The man was not joking. He was seriously concerned that a million penguins might prove to be too meagre a quantity for my purpose.

'I think I can make out with a million penguins,' I said. 'I should be able to find one or two photogenic ones among that lot. Tell me, are they all together, or scattered about?'

'Well, there are about half or three-quarters concentrated *here*,' he said, stabbing at the chart. 'And the rest are distributed all along the bay *here*.'

'Well, that seems perfect to me. Now what about somewhere to camp?'

'Ah!' said Captain Giri. 'That is the difficulty. Now, just here is the *estancia* of a friend of mine, Señor Huichi. He is not on the *estancia* at the moment. But if we went to see him he might let you stay there. It is, you see, about two kilometres from the main colony, so it would be a good place for you to stay.'

'That would be wonderful,' I said enthusiastically. 'When could we see Señor Huichi?'

The Captain consulted his watch and made a calculation.

'We can go and see him now, if you would like,' he said.

'Right!' I said, finishing my drink. 'Let's go.'

Huichi's house was on the outskirts of Deseado, and Huichi himself, when Captain Giri introduced us, was a man I took an instant liking to. Short, squat, with a weather-browned face, he had very dark hair, heavy black eyebrows and moustache, and dark brown eyes that were kind and humorous, with crow's feet at the corners. In his movements and his speech he had an air of quiet, unruffled confidence about him that was very reassuring. He stood silently while Giri explained our mission, occasionally glancing at me, as if summing me up. Then he asked a couple of questions, and, finally, to my infinite relief, he held out his hand to me and smiled broadly.

'Señor Huichi has agreed that you shall use his *estancia*,' said Giri, 'and he is going to accompany you himself, so as to show you the best places for penguins.'

'That is very kind of Señor Huichi . . . we are most grateful,' I said. 'Could we leave tomorrow afternoon, after I have seen my friend off on the plane?'

'*¿Si, si, como no?*' said Huichi when this had been translated to him. So we arranged to meet him on the morrow, after an early lunch, when we had seen Dicky off on the plane that was to take him to Buenos Aires.

So, that evening we sat in the depressing bar of our hotel, sipping our drinks and contemplating the forlorn fact that the next day Dicky would be leaving us. He had been a charming and amusing companion, who had put up with discomfort without complaint, and had enlivened our flagging spirits throughout the trip with jokes, fantastically phrased remarks, and lilting Argentine songs. We were going to miss him, and he was equally depressed at the thought of leaving us just when the trip was starting to get interesting. In a daring fit of *joie de vivre* the hotel proprietor had switched on a small radio, strategically placed on a shelf between two bottles of brandy. This now blared out a prolonged and mournful tango of the more cacophonous sort. We listened to it in silence until the last despairing howls had died away.

'What is the translation of that jolly little piece?' I asked Marie.

'It is a man who has discovered that his wife has T.B.,' she explained. 'He has lost his job and his children are starving. His wife is dying. He is very sad, and he asks the meaning of life.'

The radio launched itself into another wailing air that sounded almost identical with the first. When it had ended I raised my eyebrows inquiringly at Marie.

'That is a man who has just discovered that his wife is unfaithful,' she translated moodily. 'He has stabbed her. Now he is to be hung, and his children will be without mother or father. He is very sad and he asks the meaning of life.'

A third refrain rent the air. I looked at Marie. She listened attentively for a moment, then shrugged.

'The same,' she said tersely.

We got up in a body and went to bed.

Early the next morning Marie and I drove Dicky out to the

airstrip, while Sophie and Jacquie went round the three shops in Deseado to buy the necessary supplies for our trip out to Huichi's *estancia*. The airstrip consisted of a more or less level strip of ground on the outskirts of the town, dominated by a moth-eaten-looking hangar, whose loose boards flapped and creaked in the wind. The only living things were three ponies, grazing for-lornly. Twenty minutes after the plane had been due in there was still no sign of her, and we began to think that Dicky would have to stay with us after all. Then along the dusty road from the town came bustling a small van. It stopped by the hangar, and from inside appeared two very official-looking men in long khaki coats. They examined the wind-sock with a fine air of concentra-tion, stared up into the sky, and consulted each other with frown-ing faces. Then they looked at their watches and paced up and down.

'They must be mechanics,' said Dicky.

'They certainly look very official,' I admitted.

'Hey! Listen!' said Dicky, as a faint drone made itself heard. 'She is arrive.'

The plane came into view as a minute speck on the horizon that rapidly grew bigger and bigger. The two men in khaki coats now came into their own. With shrill cries they ran out on to the airstrip and proceeded to drive away the three ponies, who, up till then had been grazing placidly in the centre of what now turned out to be the runway. There was one exciting moment just as the plane touched down, when we thought that one of the ponies was going to break back, but one of the khaki-clad men launched himself forward and grabbed it by the mane at the last minute. The plane bumped and shuddered to a halt, and the two men left their equine charges and produced, from the depths of the hangar, a flimsy ladder on wheels which they set against the side of the plane. Apparently Dicky was the only passenger to be picked up in Deseado.

Dicky wrung my hand.

'Gerry,' he said, 'you will do for me one favour, yes?'

'Of course, Dicky,' I said, 'anything at all.'

'See that there is no bloody bastard horses in the way when we go up, eh?' he said earnestly, and then strode off to the plane, the flaps of his deer-stalker flopping to and fro in the wind.

The plane roared off, the ponies shambled back on to the runway, and we turned the blunt snout of the Land-Rover back towards the town.

We picked up Huichi at a little after twelve, and he took over the wheel of the Land-Rover. I was heartily glad of this, for we had only travelled a couple of miles from Deseado when we branched off the road on to something so vague that it could hardly be dignified with the term of track. Occasionally this would disappear altogether, and, if left to myself, I would have been utterly lost, but Huichi would aim the Land-Rover at what appeared to be an impenetrable thicket of thorn bushes, and we would tear through it, the thorns screaming along the sides of the vehicle like so many banshees, and there, on the further side, the faint wisp of track would start again. At other points the track turned into what appeared to be the three-feet-deep bed of an extinct river, exactly the same width as the Land-Rover, so we were driving cautiously along with two wheels on one bank – as it were – and two wheels on the other. Any slight miscalculation here and the vehicle could have fallen into the trough and become hopelessly stuck.

Gradually, as we got nearer and nearer to the sea, the landscape underwent a change. Instead of being flat it became gently undulating, and here and there the wind had rasped away the topsoil and exposed large areas of yellow and rust-red gravel, like sores on the furry pelt of the land. These small desert-like areas seemed to be favoured by that curious animal, the Patagonian hare, for it was always on these brilliant expanses of gravel that we found them, sometimes in pairs, sometimes in small groups of three or four. They were strange creatures, that looked as

though they had been put together rather carelessly. They had blunt, rather hare-like faces, small, neat, rabbit-shaped ears, neat forequarters with slender fore-legs. But the hindquarters were large and muscular in comparison, with powerful hind-legs. The most attractive part of their anatomy was their eyes, which were large, dark and lustrous, with a thick fringe of eyelashes. They would lie on the gravel, sunning themselves, gazing aristocratically down their blunt noses, looking like miniature Trafalgar Square lions. They would let us approach fairly close, and then suddenly their long lashes would droop over their eyes seductively, and with amazing speed they would bounce into a sitting position. They would turn their heads and gaze at us for one brief moment, and then they would launch themselves at the heat-shimmered horizon in a series of gigantic bounding leaps, as if they were on springs, the black and white pattern on their behinds showing up like a retreating target.

Presently, towards evening, the sun sank lower and in its

slanting rays the landscape took on new colours. The low growth of thorn scrub became purple, magenta and brown, and the areas of gravel were splashed with scarlet, rust, white and yellow. As we scrunched our way across one such multicoloured area of gravel we noticed a black blob in the exact centre of the expanse, and driving closer to it we discovered it was a huge tortoise, heaving himself over the hot terrain with the grim determination of a glacier. We stopped and picked him up, and the reptile, horrified by such an unexpected meeting, urinated copiously. Where he could have found, in that desiccated land, sufficient moisture to produce this lavish defensive display was a mystery. However, we christened him Ethelbert, put him in the back of the Land-Rover and drove on.

Presently, in the setting sun, the landscape heaved itself up into a series of gentle undulations, and we switchbacked over the last of these and out on to what at first looked like the level bed of an ancient lake. It lay encircled by a ring of low hills, and was, in fact, a sort of miniature dust-bowl created by the wind, which had carried the sand from the shore behind the hills and deposited it here in a thick, choking layer that had killed off the vegetation. As we roared across this flat area, spreading a fan of white dust behind us, we saw, in the lee of the further hills, a cluster of green trees, the first we had seen since leaving Deseado. As we drew nearer we could see that this little oasis of trees was surrounded by a neat white fence, and in the centre, sheltered by the trees, stood a neat wooden house, gaily painted in bright blue and white.

Huichi's two peons came to meet us, two wild-looking characters dressed in *bombachas* and tattered shirts, with long black hair and dark, flashing eyes. They helped us unload our gear and carry it into the house, and then, while we unpacked and washed, they went with Huichi to kill a sheep and prepare an *asado* in our honour. At the bottom of the slope on which the house was built, Huichi had prepared a special *asado* ground. An *asado* needs a fierce fire, and with the biting and continuous wind that blew in Patagonia you had to be careful unless you wanted to see your entire fire suddenly lifted into the air and blown away to set fire to the tinder-dry scrub for miles around. In order to guard against this Huichi had planted, at the bottom of the hill, a great square of cypress trees. These had been allowed to grow up to a height of some twelve feet, and had then had their tops lopped off, with the result that they had grown very bushy. They had been planted so close together in the first place that now their branches entwined, and formed an almost impenetrable hedge. Then Huichi had carved a narrow passage-way into the centre of this box of cypress, and had there chopped out a room, some twenty feet by twelve. This was the *asado* room, for, protected by the thick walls of cypress, you could light a fire without danger.

By the time we had washed and changed, and the sheep had been killed and stripped, it was dark; we made our way down to the *asado* room, where one of the peons had already kindled an immense fire. Near it a great stake had been stuck upright in the ground, on this a whole sheep, split open like an oyster, had been spitted. We lay on the ground around the fire and drank red wine while waiting for our meal to cook.

I have been to many *asados* in the Argentine, but that first one at Huichi's *estancia* will always remain in my mind as the most perfect. The wonderful smell of burning brushwood, mingling with the smell of roasting meat, the pink and orange tongues of flame lighting up the green cypress walls of the shelter, and the sound of the wind battering ferociously against these walls and then dying to a soft sigh as it became entangled and sapped of its strength in the mesh of branches, and above us the night sky, trembling with stars, lit by a fragile chip of moon. To gulp a mouthful of soft, warm red wine, and then to lean forward and slice a fragrant chip of meat from the brown, bubbling carcase in front of you, dunk it in the fierce sauce of vinegar, garlic and red pepper, and then stuff it, nut-sweet and juicy, into your mouth, seemed one of the most satisfying actions of my life.

Presently, when our attacks on the carcase became more desultory, Huichi took a gulp of wine, wiped his mouth with the back of his hand, and beamed at me across the red, pulsating embers of the fire, lying like a great sunset on the ground.

'¿*Mañana*,' he said, smiling, 'we go to the *pinguinos*?'

'*Si, si*,' I responded sleepily, leaning forward in sheer greed to detach another strip of crackling skin from the cooling remains of the sheep, '*mañana* the *pinguinos*.'

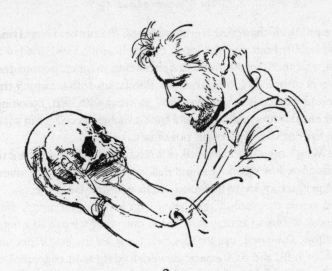

2.

A Sea of Headwaiters

It was a brave bird; and till reaching the sea, it regularly fought
and drove me backwards.

CHARLES DARWIN: THE VOYAGE OF H.M.S. BEAGLE

Early the next morning, while it was still dark, I was awoken by
Huichi moving around the kitchen, whistling softly to himself,
clattering the coffee-pot and cups, trying to break in on our slum-
bers gently. My immediate reaction was to snuggle down deeper
under the pile of soft, warm, biscuit-coloured guanaco skins that
covered the enormous double-bed in which Jacquie and I were
ensconced. Then, after a moment's meditation, I decided that if
Huichi was up I ought to be up as well; in any case, I knew I
should have to get up in order to rout the others out. So, taking a
deep breath, I threw back the bed-clothes and leapt nimbly out of
bed. I have rarely regretted an action more: it was rather like
coming freshly from a boiler-room and plunging into a mountain

stream. With chattering teeth I put on all the clothes I could find, and hobbled out into the kitchen. Huichi smiled and nodded at me, and then, in the most understanding manner, poured two fingers of brandy into a large cup, filled it up with steaming coffee and handed it to me. Presently, glowing with heat, I took off one of my three pullovers, and took a malicious delight in making the rest of the party get out of bed.

We set off eventually, full of brandy and coffee, in the pale daffodil-yellow dawn light and headed towards the place where the penguins were to be found. Knots of blank-faced sheep scuttled across the nose of the Land-Rover as we drove along, their fleeces wobbling as they ran, and at one point we passed a long, shallow dew-pond, caught in a cleft between the gentle undulation of hills, and six flamingoes were feeding at its edge, pink as cyclamen buds. We drove a quarter of an hour or so, and then Huichi swung the Land-Rover off the main track and headed across country, up a gentle slope of land. As we came to the top of the rise, he turned and grinned at me.

'Ahora,' he said, '*ahora los pinguinos.*'

Then we reached the top of the slope and there was the penguin colony.

Ahead of us the low, brown scrub petered out, and in its place was a great desert of sun-cracked sand. This was separated from the sea beyond by a crescent-shaped ridge of white sand-dunes, very steep and some two hundred feet high. It was in this desert area, protected from the sea wind by the encircling arms of the dunes, that the penguins had created their city. As far as the eye could see on every side the ground was pock-marked with nesting burrows, some a mere half-hearted scrape in the sand, some several feet deep. These craters made the place look like a small section of the moon's surface seen through a powerful telescope. In among these craters waddled the biggest collection of penguins I had ever seen, like a sea of pigmy headwaiters, solemnly shuffling to and fro as if suffering from fallen arches due to a lifetime of carrying overloaded trays. Their numbers were prodigious, stretching to the furthermost horizon where they twinkled black and white in the heat haze. It was a breath-taking sight. Slowly we drove through the scrub until we reached the edge of this gigantic honeycomb of nest-burrows and then we stopped and got out of the Land-Rover.

We stood and watched the penguins, and they stood and watched us with immense respect and interest. As long as we stayed near the vehicle they showed no fear. The greater proportion of birds were, of course, adult; but each nesting burrow contained one or two youngsters, still wearing their baby coats of down, who regarded us with big, melting dark eyes, looking rather like plump and shy debutantes clad in outsize silver-fox furs. The adults, sleek and neat in their black and white suits, had red wattles round the base of their beaks, and bright, predatory, street-pedlar eyes. As you approached them they would back towards their burrows, twisting their heads from side to side in a warning display, until sometimes they would be looking at you competely upside down. If you approached too close they would walk backwards into their burrows and gradually disappear, still twisting their heads vigorously. The babies, on the other hand,

would let you get within about four feet of them, and then their nerve would break and they would turn and dive into the burrow, so that their great fluffy behinds and frantically flapping feet were all that could be seen of them.

At first the noise and movement of the vast colony was confusing. As a background to the continuous whispering of the wind was the constant peeting of the youngsters, and the loud prolonged, donkey-like bray of the adults, standing up stiff and straight, flippers spread wide, beaks pointing at the blue sky as they brayed joyfully and exultingly. To begin with you did not know where to look first, and the constant movement of the adults and young seemed to be desultory and without purpose. Then after a few hours of getting used to being amongst such a huge assemblage of birds, a certain pattern seemed to emerge. The first thing that became obvious was that most of the movement in the colony was due to adult birds. A great number stood by the nest-burrows, obviously doing sentry duty with the young, while among them vast numbers of other birds passed to and fro, some making their way towards the sea, others coming from it. The distant sand-dunes were freckled with the tiny plodding figures of penguins, either climbing the steep slopes or sliding down them. This constant trek to and fro to the sea occupied a large portion of the penguins' day, and it was such a tremendous feat that it deserves to be described in detail. By carefully watching the colony, day by day, during the three weeks we lived among it, we discovered that this is what happened:

Early in the morning one of the parent birds (either male or female) would set out towards the sea, leaving its mate in charge of the nestlings. In order to get to the sea the bird had to cover about a mile and a half of the most gruelling and difficult terrain imaginable. First they had to pick their way through the vast patchwork of nesting burrows that made up the colony, and when they reached the edge of this – the suburbs, as it were – they were faced by the desert area, where the sand was caked and

split by the sun into something resembling a gigantic jig-saw puzzle. The sand in this area would, quite early in the day, get so hot that it was painful to touch, and yet the penguins would plod dutifully across it, pausing frequently for a rest, as though in a trance. This used to take them about half an hour. But, when they reached the other side of the desert they were faced with another obstacle, the sand-dunes. These towered over the diminutive figures of the birds like a snow-white chain of Himalayan mountains, two hundred feet high, their steep sides composed of fine, loose shifting sand. We found it difficult enough to negotiate these dunes, so it must have been far worse for such an ill-equipped bird as a penguin.

When they reached the base of the dunes they generally paused for about ten minutes to have a rest. Some just sat there, brooding, while others fell forwards on to their tummies and lay there panting. Then, when they had rested, they would climb sturdily to their feet and start the ascent. Gathering themselves, they would rush at the slope, obviously hoping to get the worst of the climb over as quickly as possible. But this rapid climb would peter out about a quarter of the way up; their progress would slow down, and they would pause to rest more often. As the gradient grew steeper and steeper they would eventually be forced to flop down on their bellies, and tackle the slope that way, using their flippers to assist them in the climb. Then, with one final, furious burst of speed, they would triumphantly reach the top, where they would stand up straight, flap their flippers in delight, and then flop down on to their tummies for a ten-minute rest. They had reached the half-way mark and, lying there on the knife-edge top of the dune, they would see the sea, half a mile away, gleaming coolly and enticingly. But they had still to descend the other side of the dune, cross a quarter of a mile of scrub-land and then several hundred yards of shingle beach before they reached the sea.

Going down the dune, of course, presented no problem to them, and they accomplished this in two ways, both equally amusing to watch. Either they would walk down, starting very sedately and getting quicker and quicker the steeper the slope became, until they were galloping along in the most undignified way, or else they would slide down on their tummies, using their wings and feet to propel their bodies over the surface of the sand exactly as if they were swimming. With either method they reached the bottom of the dune in a small avalanche of fine sand, and they would get to their feet, shake themselves, and set off grimly through the scrub towards the beach. But it was the last few hundred yards of beach that seemed to make them suffer most. There was the sea, blue, glittering, lisping seductively on the shore, and to get to it they had to drag their tired bodies over the stony beach, where the pebbles scrunched and wobbled under their feet, throwing them off balance. But at last it was over, and they ran the last few feet to the edge of the waves in a curious crouching position, then suddenly straightened up and plunged into the cool water. For ten minutes or so they twirled and ducked in a shimmer of sun ripples, washing the dust and sand from their heads and wings, fluttering their hot, sore feet in the water in ecstasy, whirling and bobbing, disappearing beneath

the water, and popping up again like corks. Then, thoroughly refreshed, they would set about the stern task of fishing, undaunted by the fact that they would have to face that difficult journey once again before the food they caught could be delivered to their hungry young.

Once they had plodded their way – full of fish – back over the hot terrain to the colony, they would have to start on the hectic job of feeding their ravenous young. This feat resembled a cross between a boxing- and an all-in wrestling-match, and was fascinating and amusing to watch. There was one family that lived in a burrow close to the spot where we parked the Land-Rover each day, and both the parent birds and their young got so used to our presence that they allowed us to sit and film them at a distance of about twenty feet, so we could see every detail of the feeding process very clearly. Once the parent bird reached the edge of the colony it had run the gauntlet of several thousand youngsters before it reached its own nest-burrow and babies. All these youngsters were convinced that, by launching themselves at the adult bird in a sort of tackle, they could get it to regurgitate the food it was carrying. So the adult had to avoid the attacks of these fat, furry youngsters by dodging to and fro like a skilful centre-forward on a football field. Generally the parent would end up at its nest-burrow, still hotly pursued by two or three strange chicks, who were grimly determined to make it produce food. When it reached home the adult would suddenly lose patience with its pursuers, and, rounding on them, would proceed to beat them up in no uncertain fashion, pecking at them so viciously that large quantities of the babies' fluff would be pecked away, and float like thistledown across the colony.

Having routed the strange babies, it would then turn its attention to its own chicks, who were by now attacking it in the same way as the others had done, uttering shrill wheezing cries of hunger and impatience. It would squat down at the entrance to the burrow and stare at its feet pensively, making motions like someone trying to stifle an acute attack of hiccups. On seeing this the youngsters would work themselves into a frenzy of delighted anticipation, uttering their wild, wheezing cries, flapping their wings frantically, pressing themselves close to the parent bird's body, and stretching up their beaks and clattering them against the adult's. This would go on for perhaps thirty seconds, when the parent would suddenly – with an expression of relief – regurgitate vigorously, plunging its beak so deeply into the gaping mouths of the youngsters that you felt sure it would never be able to pull its head out again. The babies, satisfied and apparently not stabbed from stem to stern by the delivery of the first course, would squat down on their plump behinds and meditate for a while, and their parent would seize the opportunity to have a quick wash and brush up, carefully preening its breast-feathers, picking minute pieces of dirt off its feet, and running its beak along its wings with a clipper-like motion. Then it would yawn, bending forward like someone attempting to touch his toes, wings stretched out straight behind, beak gaping wide. Then it would sink into the trance-like state that its babies had attained some minutes earlier. All would be quiet for five minutes or so, and then suddenly the parent would start its strange hiccupping motions again, and pandemonium would break out immediately. The babies would rouse themselves from their digestive reverie and hurl themselves at the adult, each trying its best to get its beak into position first. Once more each of them in turn would be apparently stabbed to the heart by the parent's beak, and then once more they would sink back into somnolence.

The parents and young who occupied this nest-burrow where we filmed the feeding process were known, for convenient reference, as the Joneses. Quite close to the Joneses' establishment was another burrow that contained a single, small and very undernourished-looking chick whom we called Henrietta Vacanttum. Henrietta was the product of an unhappy home-life. Her parents were, I suspected, either dimwitted or just plain idle, for they took twice as long as any other penguins to produce food for Henrietta, and then only in such minute quantities that she was always hungry. An indication of her parents' habits was the slovenly nest-burrow, a mere half-hearted scrape, scarcely deep enough to protect Henrietta from any inclement weather, totally unlike the deep, carefully dug villa-residence of the Jones family. So it was not surprising that Henrietta had a big-eyed, half-starved, ill-cared-for look about her that made us feel very sorry for her. She was always on the look-out for food, and as the Jones parents had to pass her front door on their way to their own neat burrow, she always made valiant attempts to get them to regurgitate before they reached home.

These efforts were generally in vain, and all Henrietta got for her pains was a severe pecking that made her fluff come out in great clouds. She would retreat, disgruntled, and with anguished eye watch the two disgustingly fat Jones babies wolfing down their food. But one day, by accident, Henrietta discovered a way

to pinch the Jones family's food without any unpleasant repercussions. She would wait until the parent Jones had started the hiccupping movements as a preliminary to regurgitation, and the baby Joneses were frantically gyrating round, flapping their wings and wheezing, and then, at the crucial moment, she would join the group, carefully approaching the parent bird from behind. Then, wheezing loudly, and opening her beak wide, she would thrust her head either over the adult's shoulder, as it were, or under its wing, but still carefully maintaining her position behind the parent so that she should not be recognized. The parent Jones, being harried by its gaping-mouthed brood, its mind fully occupied with the task of regurgitating a pint of shrimps, did not seem to notice the introduction of a third head into the general mêlée that was going on around it. And when the final moment came it would plunge its head into the first gaping beak that was presented, with the slightly desperate air of an aeroplane passenger seizing his little brown paper bag at the onset of the fiftieth air-pocket. Only when the last spasm had died away, and the parent Jones could concentrate on external matters, would it realize that it had been feeding a strange offspring, and then Henrietta had to be pretty nifty on her great, flat feet to escape the wrath. But even if she did not move quickly enough, and received a beating up for her iniquity, the smug look on her face seemed to argue that it was worth it.

In the days when Darwin had visited this area there had still been the remnants of the Patagonian Indian tribes left, fighting a losing battle against extermination by the settlers and soldiers. These Indians were described as being uncouth and uncivilized and generally lacking in any quality that would qualify them for a little Christian charity. So they vanished, like so many animal species when they come into contact with the beneficial influences of civilization, and no one, apparently, mourned their going. In various museums up and down Argentina you can see a few remains of their crafts (spears, arrows, and so on) and inevitably a large and rather gloomy picture purporting to depict the more unpleasant side of the Indians' character, their lechery. In every one of these pictures there was shown a group of long-haired, wild-looking Indians on prancing wild steeds, and the leader of the troupe inevitably had clasped across his saddle a white woman in a diaphanous garment, whose mammary development would give any modern film star pause for thought. In every museum the picture was almost the same, varying only in the number of Indians shown, and the chest expansion of their victim. Fascinating though these pictures were, the thing that puzzled me was that there was never a companion piece to show a group of civilized white men galloping off with a voluptuous Indian girl, and yet this had happened as frequently (if not more frequently) than the rape of white women. It was a curious and interesting sidelight on history. But nevertheless these spirited but badly painted portraits of abduction had one interesting feature. They were obviously out to give the worst possible impression of the Indians, and yet all they succeeded in doing was in impressing you with a wild and rather beautiful people, and filling you with a pang of sorrow that they were no longer in existence. So, when we got down into Patagonia I searched eagerly for relics of these Indians, and questioned everyone for stories about them. The stories, unfortunately, were much of a muchness and told me little, but when it came to relics, it turned out, I could not have gone to a better place than the penguin metropolis.

One evening, when we had returned to the *estancia* after a hard day's filming and were drinking *maté* round the fire, I asked Señor Huichi – *via* Marie – if there had been many Indian tribes living in those parts. I phrased my questions delicately, for I had been told that Huichi had Indian blood in him, and I was not sure whether this was a thing he was proud of or not. He smiled his slow and gentle smile, and said that on and around his *estancia* had been one of the largest concentrations of Indians in Patagonia. In fact, he went on, the place where the penguins lived still yielded evidence of their existence. What sort of evidence, I asked eagerly. Huichi smiled again, and, getting to his feet he disappeared into his darkened bedroom. I heard him pull a box out from under his bed, and he returned carrying it in his hands and placed it on the table. He removed the lid and tipped the contents out on to the white tablecloth, and I gasped.

I had seen, as I say, various relics in the museums, but nothing to compare with this; for Huichi tumbled out on to the table a rainbow-coloured heap of stone objects that were breath-taking in their colouring and beauty. There were arrowheads ranging from delicate, fragile-looking ones the size of your little finger-nail, to ones the size of an egg. There were spoons made by slicing in half and carefully filing down big sea-shells; there were long, curved stone scoops for removing the edible molluscs from their shells; there were spearheads with razor-sharp edges; there were the balls for the *boleadoras*, round as billiard-balls, with a shallow trough running round their equators, as it were, which took the thong from which they hung; these were so incredibly perfect that one could hardly believe that such precision could be achieved without a machine. Then there were the purely decorative articles: the shells neatly pierced for ear-rings, the necklace made of beautifully matched green, milky stone rather like jade, the seal-bone that had been chipped and carved into a knife that was obviously more ornamental than useful. The pattern on it was simple arrangements of lines, but carved with great precision.

I sat poring over these objects delightedly. Some of the arrowheads were so small it seemed impossible that anyone could create them by crude chipping, but hold them up to the light and you could see where the delicate wafers of stone had been chipped away. What was more incredible still was that each of these arrowheads, however small, had a minutely serrated edge to give it a bite and sharpness. As I was examining the articles I was suddenly struck by their colouring. On the beaches near the penguins almost all the stones were brown or black; to find attractively coloured ones you had to search. And yet every arrowhead, however small, every spearhead, in fact every piece of stone that had been used had obviously been picked for its beauty. I arranged all the spear- and arrowheads in rows on the tablecloth, and they lay there gleaming like the delicate leaves from some fabulous tree. There were red ones with a darker vein of red, like dried blood; there were green ones covered with a fine tracery of white; there were blue-white ones, like mother-of-pearl; and yellow and white ones covered with a freckling of blurred patterns in blue or black where the earth's juices had stained the stone. Each piece was a work of art, beautifully shaped, carefully and minutely chipped, edged and polished, constructed out of the most beautiful piece of stone the maker could find. You could see they had been made with love. And these, I reminded myself, were made by the barbarous, uncouth, savage and utterly uncivilized Indians for whose passing no one appeared to be sorry.

Huichi seemed delighted that I should display such obvious interest and admiration for his relics, and he went back into the bedroom and unearthed another box. This one contained an extraordinary weapon carved from stone: it was like a small dumb-bell. The central shaft which connected the two great, misshapen balls of stone fitted easily into the palm of your hand, so that then you had a great ball of stone above and below your fist. As the whole thing weighed about three pounds it was a fearsome weapon, capable of splitting a man's skull like a puffball. The next item in the box – which Huichi reverently unwrapped from a sheet of tissue-paper – looked as though, in fact, it had been treated with this stone club. It was an Indian skull, white as ivory, with a great splinter-edged gaping hole across the top of the cranium.

Huichi explained that over the years, whenever his work had taken him to the corner of the *estancia* where the penguins lived, he had searched for Indian relics. He said that the Indians had apparently used that area very extensively, for what particular purpose no one was quite sure. His theory was that they had used the great flat area where the penguins now nested as a sort of arena, where the young men of the tribe practised shooting with bow and arrow, spear-throwing, and the art of entangling their quarry's legs with the *boleadoras*. On the other side of the great sand-dunes, he said, were to be found huge piles of empty sea-shells. I had noticed these great, white heaps of shells, some covering an area of a quarter of an acre and about three feet thick, but I had been so engrossed in my filming of the penguins that I had only given them a passing thought. Huichi's theory was that this had been a sort of holiday resort, as it were, the Margate of the Indians. They had come down there to feed on the succulent and plentiful shellfish, to find stones on the shingle beach from which to make their weapons, and a nice flat area on which to practise with these weapons. What other reason would there be for finding these great piles of empty shells, and, scattered over the sand-dunes and shingle patches, such a host of arrow- and spearheads, broken necklaces, and the occasional crushed skull? I must say Huichi's idea seemed to me to be a sensible one, though I suppose a professional archaeologist would have found some method of disproving it. I was horrified at the thought of the number of delicate and lovely arrowheads that must have been splintered and crushed beneath the Land-Rover wheels as we had gaily driven to and fro over the penguin town. I resolved that the next day, when we had finished filming, we would search for arrowheads.

As it happened, the next day we had only about two hours' decent sunshine suitable for filming, and so the rest of the time we spent crawling over the sand-dunes in curious prenatal postures, searching for arrowheads and other Indian left-overs. I very soon discovered that it was not nearly as easy as it seemed. Huichi, after years of practice, could spot things with uncanny accuracy from a great distance.

'*Esto, una,*' he would say, smiling, pointing with the toe of his shoe at a huge pile of shingle. I would glare at the area indicated, but could see nothing but unworked bits of rock.

'*Esto*,' he would say again, and bending down pick up a beautiful leaf-shaped arrowhead that had been within five inches of my hand. Once it had been pointed out, of course, it became so obvious that you wondered how you had missed it. Gradually, during the course of the day, we improved, and our pile of finds started mounting, but Huichi still took a mischievous delight in wandering erect behind me as I crawled laboriously across the dunes, and, as soon as I thought I had sifted an area thoroughly, he would stoop down and find three arrowheads which I had somehow missed. This happened with such monotonous regularity that I began to wonder, under the influence of an aching back and eyes full of sand, whether he was not palming the arrowheads, like a conjuror, and pretending to find them just to pull my leg. But then my unkind doubts were dispelled, for he suddenly leant forward and pointed at an area of shingle I was working over.

'*Esto*,' he said, and, leaning down, pointed out to me a minute area of yellow stone protruding from under a pile of shingle. I gazed at it unbelievingly. Then I took it gently between my fingers and eased from under the shingle a superb yellow arrowhead with a meticulously serrated edge. There had been approximately a quarter of an inch of the side of the arrowhead showing, and yet Huichi had spotted it.

However, it was not long before I got my own back on him. I was making my way over a sand-dune towards the next patch of shingle, when my toe scuffed up something that gleamed white. I bent down and picked it up, and to my astonishment found I was holding a beautiful harpoon-head about six inches long, magnificently carved out of fur seal bone. I called to Huichi, and when he saw what I had found his eyes widened. He took it from me gently and wiped the sand off it, and then turned it over and over in his hands, smiling with delight. He explained that a harpoon-head like this was one of the rarest things you could find. He had only ever found one, and that had been so crushed that it had not been worth saving. Ever since he had been looking, without success, for a perfect one to add to his collection.

Presently it was getting towards evening, and we were all scattered about the sand-dunes hunched and absorbed in our task. I rounded a spur of sand and found myself in a tiny valley between the high dunes, a valley decorated with two or three wizened and carunculated trees. I paused to light a cigarette and ease my aching back. The sky was turning pink and green as it got towards sunset time, and apart from the faint whisper of the sea and the wind it was silent and peaceful. I walked slowly up the little valley, and suddenly I noticed a slight movement ahead of me. A small, very hairy armadillo was scuttling along the top of the dunes like a clockwork toy, intent on his evening search for food. I watched him until he disappeared over the dunes and then walked on. Under one of the bushes I was surprised to see a pair of penguins, for they did not usually choose this fine sand to dig their nest-burrows in. But this pair had chosen this valley for some reason of their own, and had scraped and scrabbled a rough hole in which squatted a single fur-coated chick. The parents castanetted their beaks at me and twisted their heads upside down, very indignant that I should disturb their solitude. I watched them for a moment, and then I noticed something half hidden in the pile of sand which they had dug out to form their nest. It was something smooth and white. I went forward and, despite the near hysterics of the penguins, I scraped away the sand. There lying in front of me was a perfect Indian skull, which the birds must have unearthed.

I sat down with the skull on my knee and smoked another cigarette while I contemplated it. I wondered what sort of a man this vanished Indian had been. I could imagine him, squatting on the shore, carefully and cleverly chipping minute flakes off a piece of stone to make one of the lovely arrowheads that now squeaked and chuckled in my pocket. I could imagine him, with his fine brown face and dark eyes, his hair hanging to his shoulders, his rich brown guanaco skin cloak pulled tight about him as he sat very straight on a wild, unshod horse. I gazed into the empty eye-sockets of the skull and wished fervently that I could have met the man who had produced anything as beautiful as those arrowheads. I wondered if I ought to take the skull back to England with me and give it a place of honour in my study, surrounded by his artistic products. But then I looked around, and decided against it. The sky was now a vivid dying blue, with pink and green thumb-smudges of cloud. The wind made the sand trickle down in tiny rivulets that hissed gently. The strange, witch-like bushes creaked pleasantly and musically. I felt that the Indian would not mind sharing his last resting place with the creatures of what had once been his country, the penguins and the armadillos. So I dug a hole in the sand and placing the skull in it I gently covered it over. When I stood up in the rapidly gathering gloom the whole area seemed steeped in sadness, and the presence of the vanished Indians seemed very close. I could almost believe that, if I looked over my shoulder quickly, I would see one on horseback, silhouetted against the coloured sky. I shrugged this feeling off as fanciful, and walked back towards the Land-Rover.

As we rattled and bumped our way back in the dusk towards the *estancia*, Huichi, talking to Marie, said very quietly:

'You know, señorita, that place always seems to be sad. I feel the Indians there very much. They are all around you, their ghosts, and one feels sorry for them because they do not seem to be happy ghosts.'

This had been my feeling exactly.

Before we left the next day I gave Huichi the harpoon-head I had found. It broke my heart to part with it, but he had done so much for us that it seemed very small return for his kindness. He was delighted, and I know that it is now reverently wrapped in tissue-paper in the box beneath his bed, not too far from where it ought to be, buried on the great shining dunes, feeling only the shifting sand as the penguins thump solidly overhead.

3.

The Golden Swarm

They appeared to be of a loving disposition, and lay huddled
together, fast asleep, like so many pigs.

CHARLES DARWIN: THE VOYAGE OF H.M.S. BEAGLE

The penguin colony near Huichi's *estancia* had been our southern-
most goal. Now, leaving Deseado behind us we drove northward
across the flat purple scrub-land towards Peninsula Valdes, where,
I had been assured, I would find large colonies of fur seals, and
the only remaining colony of elephant seals in Argentina.

Peninsula Valdes lies on the coast of the province of Chubut.
It is a mass of land rather like an axe-head, some eighty miles
long by thirty broad. The peninsula is almost an island, being
connected to the mainland by such a narrow neck of land that, as
you drive along it, you can see the sea on both sides of the road.
Entering the peninsula was like coming into a new land. For days
we had driven through the monotonous and monochrome

Patagonian landscape, flat as a billiard-table and apparently devoid of life. Now we reached the fine neck of land on the other side of which was the peninsula, and suddenly the landscape changed. Instead of the small, spiky bushes stretching purply to the horizon, we drove into a buttercup-yellow landscape, for the bushes were larger, greener and each decked with a mass of tiny blooms. The countryside was no longer flat but gently undulating, stretching away to the horizon like a yellow sea, shimmering in the sun.

Not only had the landscape changed in colouring and mood but it had suddenly become alive. We were driving down the red earth road, liberally sprinkled with backbreaking potholes, when suddenly I caught a flash of movement in the undergrowth at the side of the road. Tearing my eyes away from the potholes I glanced to the right, and immediately trod on the brakes so fiercely that there were frenzied protests from all the female members of the party. But I simply pointed, and they became silent.

To one side of the road, standing knee-deep in the yellow bushes, stood a herd of six guanacos, watching us with an air of intelligent interest. Now guanacos are wild relatives of the llama, and I had been expecting to see something that was the same rather stocky shape as the llama, with a dirty brown coat. At least, I remembered that the one I had seen in a zoo many years before looked like that. But either my memory had played me false or else it had been a singularly depressed specimen I had seen. It had certainly left me totally unprepared for the magnificent sight these wild guanacos made.

What I took to be the male of the herd was standing a little in front of the others and about thirty feet away from us. He had long, slender racehorse legs, a streamlined body and a long graceful neck reminiscent of a giraffe's. His face was much longer and more slender than a llama's, but wearing the same supercilious expression. His eyes were dark and enormous. His small neat ears twitched to and fro as he put up his chin and examined us as if through a pair of imaginary lorgnettes. Behind him, in a tight and timid bunch, stood his three wives and two babies, each about the size of a terrier, and they had such a look of wide-eyed innocence that it evoked strange anthropomorphic gurgles and gasps from the feminine members of the expedition. Instead of the dingy brown I had expected these animals almost glowed. The neck and legs were a bright yellowish colour, the colour of sunshine on sand, while their bodies were covered with a thick fleece of the richest biscuit brown. Thinking that we might not get such a chance again I determined to get out of the Land-Rover and film them. Grabbing the camera I opened the door very slowly and gently. The male guanaco put both ears forward and examined my manoeuvre with manifest suspicion. Slowly I closed the door of the Land-Rover and then started to lift the camera. But this was enough. They did not mind my getting out of the vehicle, but when I started to lift a black object – looking suspiciously like a gun – to my shoulder this was more than they could stand. The male uttered a snort, wheeled about, and galloped off, herding his females and babies in front of him. The babies were inclined to think this was rather a lark, and started gambolling in circles, until their father called them to order with a few well-directed kicks. When they got some little distance away they slowed down from their first wild gallop into a sedate, stiff-legged canter. They looked, with their russet and yellow coats, like some strange gingerbread animals, mounted on rockers, tipping and tilting their way through the golden scrub.

As we drove on across the peninsula we saw many more groups of guanacos, generally in bunches of three or four, but once we saw a group of them standing on a hill, outlined against a blue sky, and I counted eight individuals in the herd. I noticed that the herds were commoner towards the centre of the peninsula, and became considerably less common as you drove towards the coast. But wherever you saw them they were cautious and

nervous beasts, ready to canter off at the faintest hint of anything unusual, for they are persecuted by the local sheep-farmers, and have learnt from bitter experience that discretion is the better part of valour.

By the late afternoon we were nearing Punta del Norte on the east coast of the peninsula, and the road had faded away into a pair of faint wheel-tracks that wended their way through the scrub in a looping and vague manner that made me doubt whether they actually led anywhere. But, just when I was beginning to think that we had taken the wrong track, I saw up ahead a small white *estancia*, its shutters tightly fastened, and to the left of it a large Dutch barn or *galpón*. Knowing that a *galpón* was generally the centre of any activity on an *estancia*, I drove up to it and stopped. Three large, fat dogs immediately appeared, barked at us vigorously, and then, obviously thinking that their duty was done, set about the fascinating task of irrigating the Land-Rover wheels. Three peons came out from inside the barn, brown, lean, rather wild-looking men with wide, eager smiles. They were obviously delighted to see us, for strangers there were a rarity. They insisted that we go into the barn, brought chairs for us to sit on, and within half an hour they had killed a sheep and an *asado* was being prepared, while we sat and drank wine and told them why we had come.

They were fascinated by the thought that I should have come all the way from England just to catch and film *bichos*, and doubtless thought I was more than a little mad, though they were far too well-mannered to say so. On the subject of elephant seals and fur seals they were very informative and helpful. The elephant seals, they explained, had now had their babies and reared them. This meant that they were no longer to be found in one spot on the beach near the fur seals, which acted, as it were, as their maternity ward. Now they drifted up and down the coast as the mood took them, and were difficult to find, though there were two or three places which they were particularly fond of where they might be located. These favourite haunts were called, charmingly enough, the *elefanterías*. The peons marked on the map the areas in which the *elefanterías* were to be found, and then they showed me where the biggest concentration of fur seals lived. These, they said, would be easy, for they still had young, and were therefore packed on the beach and easily accessible. Moreover, the peons went on, there was a good camping area just near the fur seal colony, a flat grassy space, sheltered from the wind on all sides by a gentle rise in the ground. Cheered by this news we drank more wine, ate large quantities of roast sheep, and then clambered into the Land-Rover again and set off to look for the camp site.

We found it without too much difficulty, and it was as good as the peons had promised, a small, level plain covered with coarse grass and occasional clumps of small, twisted dead bushes. On three sides it was protected by a curving rim of low hills, covered in yellow bushes, and on the third side a high wall of shingle lay between it and the sea. This offered us some cover, but even so there was a strong and persistent wind blowing from the sea, and now that it was evening it became very cold. It was decided that the three female members of the party would sleep inside the Land-Rover, while I slept under it. Then we dug a hole, collected dry brushwood and built a fire to make tea. One had to be very careful about the fire, for we were surrounded by acres and acres of tinder-dry undergrowth, and the strong wind would, if you were not careful, lift your whole fire up into the air and dump it down among the bushes. I dreaded to think what the ensuing conflagration would be like.

The sun set in a nest of pink, scarlet and black clouds, and there was a brief green twilight. Then it darkened, and a huge yellow moon appeared and gazed down at us as we crouched around the fire, huddled in all the clothes we could put on, for the wind was now bitter. Presently the Land-Rover party crept inside the vehicle, with much grunting and argument as to whose feet should go where, and I collected my three blankets, put earth on the fire, and then fashioned myself a bed under the back axle of the Land-Rover. In spite of the fact that I was wearing three pullovers, two pairs of trousers, a duffel-coat and a woolly hat, and had three blankets wrapped round me, I was still cold, and as I shivered my way into a half-sleep I made a mental note that on the morrow I would reorganize our sleeping arrangements.

I awoke in that dimly lit silence just before dawn, when even the sound of the sea seems to have hushed. The wind had switched direction in the night, and the wheels of the Land-Rover now offered no protection at all. The hills around were black against the blue-green of the dawn sky, and there was no sound except the hiss of the wind and the faint snore of the surf. I lay there, shuddering in my cocoon of clothes and blankets, and debated whether or not I should get up and light the fire and make some tea. Cold though I was under my clothes, it was still a few degrees warmer than wandering about collecting brushwood, and so I decided to stay where I was. I was just trying to insinuate my hand into my duffel-coat pocket for my cigarettes, without letting a howling wind into my cocoon of semi-warmth, when I realized that we had a visitor.

Suddenly a guanaco stood before me, as if conjured out of nothing. He stood some twenty feet away, quite still, surveyed me with a look of surprise and displeasure, his neat ears twitching back and forth. He turned his head, sniffing the breeze, and I could see his profile against the sky. He wore the supercilious expression of his race, the faint aristocratic sneer, as if he knew that I had slept in my clothes for the past three nights. He lifted one forefoot daintily, and peered down at me closely. Whether, at that moment, the breeze carried my scent to him I don't know, but he suddenly stiffened and, after a pause for meditation, he belched.

It was not an accidental gurk, the minute breach of good manners that we are all liable to at times. This was a premeditated, rich and prolonged belch, with all the fervour of the Orient in it. He paused for a moment, glaring at me, to make sure that his comment on my worth had made me feel properly humble, and then he turned and disappeared as suddenly as he had come, and I could hear the faint whisper of his legs brushing through the little bushes. I waited for a time to see if he would come back, but he had obviously gone about his business, so I lit my cigarette and lay shivering and smoking until the sun came up.

Once we had breakfasted and everyone was more or less conscious, we unhitched the trailer, removed all our equipment from inside the Land-Rover and piled it on the ground under tarpaulins, checked the camera equipment, made sandwiches and coffee, and then set off to look for the fur seals. The peons had told us that if we drove half a mile or so down the track and then branched off, across country, towards the sea, we should easily find the colony. What they had not told us, of course, was that driving across country was a nerve- and spine-shattering experience, for the ground was corrugated and pitted in the most extraordinary way, and most of these death-traps were concealed by the bushes, so you would crash into them before you knew they were there, while the bushes screeched along the sides of the Land-Rover in what sounded like an ecstasy of shrill, maniacal laughter. At last I decided that, unless we wanted a broken spring or puncture, we had better continue the hunt on foot, so, finding a more or less level piece of country I parked the Land-Rover and we got out. At once I became aware of a strange sound, like the frenzied roar of a football crowd heard distantly. We walked through waist-high golden scrub until we came out on the edge of a small cliff, and there on the shingle beach below us, at the edge of the creaming waves, lay the fur seal colony.

As we reached this vantage point the noise of the animals smote us, roar, bleat, gurgle and cough, a constant undulation of sound, like the boiling of an enormous cauldron of porridge. The colony, consisting of about seven hundred animals, lay strung out along the beach in a line some ten or twelve deep, and so tightly packed together that, as they shifted and moved in the sun, they gleamed gold, like a restless swarm of bees. Forgetting all about filming I just squatted on the edge of the cliff, staring down at this wonderful collection of animals, completely entranced.

At first we found – as we had done with the penguin colony – that there was so much going on, so much confusion and noise, that you were bewildered, and your eyes were moving constantly up and down this immense moving plate of animals in an effort to catch and translate every movement, until you began to feel dizzy. But, after the first hour, when the shock of seeing such a magnificent mass of animals at close range had worn off somewhat, you found you could concentrate.

It was the adult bulls that first caught and held your attention, for they were so massive. They were quite the most proud and extraordinary-looking animals I have ever seen. They sat with their faces pointed skywards, their shaggy necks bent back so that the fat was scalloped into folds, their snub-noses and fat beery faces peering up into the sky with all the pompous arrogance of the Tenniel illustration of Humpty Dumpty. They had physiques like boxers, the tremendous muscular shoulders tapering down to slender hindquarters, and ending, incongruously, in a pair of limbs that were quite ridiculous. The feet had long slender fingers, carefully webbed, so the impression was that the seal was wearing, for some reason best known to himself, a pair of very elegant frogmen's flippers. Sometimes you would see one old bull stretched out asleep in the sand, blubbering and snoring to himself, while at the end of his body he would be waving his large flippers to and fro, pointing the slender fingers with all the grace and delicacy of a Balinese dancer. When they walked these huge frog-like feet stuck out on either side, and, as the motion of the animal's body was very like a rumba, the effect was extremely funny. Their colouring ranged from chocolate to a rich biscuit brown, fading to russet on the shaggy fur round their shoulders and necks. This made a nice contrast to the wives who were very much smaller and decked out in silver or golden coats. Whereas their husbands were enormous blundering tanks of animals, the wives were slim, sinuous and sexy, with their neat pointed faces and big melting eyes. They were the personification of femininity, graceful to a degree, beautiful, coquettish and at the same time loving. They were heavenly creatures, and I decided that should I ever have the chance of being an animal in this world I would choose to be a fur seal so that I might enjoy having such a wonderful wife.

Although they had some six miles of beach to use, the colony chose to lie in a tight conglomeration, covering an area about a quarter of a mile in length. It seemed to me that if they had spaced themselves out a bit more they would have halved the troubles of the colony, for, packed tightly like this, each bull was in a constant state of nerves over his little group of wives, and throughout the colony there were fights breaking out all the time. A lot of the blame for these, I am afraid, was due to the females who – as soon as they thought their husband was not watching – would undulate gracefully across the sand towards the next group, and sit there watching the bull with languishing eyes. It would take a very staunch Presbyterian fur seal to resist the appeal of those pleading melting eyes. But before any infidelity could take place the husband would suddenly make a rapid count and discover that he was a wife short. As soon as he spotted her, he would surge after her, his enormous bulk scattering the shingle like spray, and from his mouth, with its great white fangs, would issue a prolonged, lion-like belching roar. Reaching her he would catch her by the scruff of the neck and shake her savagely from side to side. Then, with a jerk of his head, he would send her spinning across the sand towards his harem.

By this time the other bull would have worked himself into a state of nerves. He would feel that the husband was too close to *his* wives for safety, and so he would lunge forward with open mouth, uttering fearsome gurgling cries, and the two would join in battle. Most of these fights were merely mock combats, and after a good deal of mouth-opening, roaring, and lunging, honour would be satisfied. But occasionally both bulls would lose their tempers, and then it was incredible and frightening to watch how two such ponderous and dropsical-looking creatures could turn into such swift, deft and deadly fighters. The shingle would be churned up as the two colossal creatures snapped and barged at each other's fat necks, and the blood spurted out over the fascinated audience of wives and babies. One of the favourite gambits during these fights was to undulate across the shingle towards your opponent, waving your head from side to side, like a boxer feinting. Then, when you got near enough you would lunge forward and, with a sideways and downwards bite, try and slash open the thick hide of your antagonist's neck. Most of the old bulls on the beach had fresh wounds or white scars decorating their necks, and one I saw looked as though someone had slashed

him with a sabre, for the wound was some eighteen inches long and appeared to be about six inches deep.

When a bull waddled back to his wives after such a battle they would gather round him in admiration and love, elongating their sinuous necks so that they could reach up and nuzzle and kiss his face, rubbing their gold and silver bodies against his barrel chest, while he stared up into the sky arrogantly, occasionally condescending to bend his head and bite one of his wives gently on the neck.

A lot of the nervous tension that the bulls with wives suffered from, and a lot of the actual fighting, was due to the bachelor bulls. These were gay young bulls, much slimmer and less muscular than the old ones, who had been unable to acquire a wife or wives for themselves at the beginning of the breeding season when the courtship battles take place. These young bulls spent most of their time just sleeping in the sun, or swimming about in the shallow water at the sea's edge. But, every now and then, they would be smitten with an impish desire to irritate their elders and betters. They would swagger slowly along the colony, their great frog's feet stuck out, gazing about them with a benign air of innocence, as though there was not an evil thought in their heads. Then, as they passed a family group in the centre of which squatted an old bull star-gazing, the young bachelor would suddenly swerve and break into an undulating run, getting faster and faster as he approached the group. The females would scatter wildly as he burst through their circle, he would hurl himself at the old bull, give him a quick bite on the neck, and then undulate rapidly away before the old bull really knew what was happening. Then, with a roar of rage the old bull would give chase, but by then the gay bachelor had reached the sea and plunged in, so the old bull, grumbling to himself, would return to round up his scattered wives, and settle himself in their midst for another period of astronomical research.

The ones that seemed to lead the most carefree and pleasant lives were the young, but fully adult bulls, who had only succeeded in getting themselves one wife. They generally lay a little apart from the main colony, their wife and cub alongside them, and spent a lot of time sleeping. They could afford to do this, as it was obviously easier to control one of these high-spirited female seals than to try and cope with the vagaries of six or seven. It was with one of these young newlywed couples that I was lucky enough to see the consummation of their marriage, as it were, and I have never seen such a delicate and beautiful piece of love-play between two animals.

The young bull had dug himself the fur seal equivalent of a honeymoon cottage in the shingle near the base of the cliff from which I was filming. This cottage consisted of a large, deep hole scraped out with his fore-flippers, so that the top layer of sun-heated shingle was scraped off, and the cool damp shingle beneath was exposed. He lay in this hole with his wife in a very typical attitude, his great head resting on her back as she lay asleep, at right-angles to him. They had lain like this, almost unmoving, for the whole morning. Now, at mid-day with the fierce sun directly overhead, they began to get restless. The bull started to wave his hind flipper to and fro in the air, shift his bulk about uneasily, and scoop great flippersful of damp shingle and shovel them on to his back, in an effort to keep cool. His wife, disturbed by his movements, woke up, looked about her, yawned widely, and then lay down again with a deep, contented sigh, gazing around placidly with her great, dark eyes. After a few minutes' contemplation she shifted her body round so that she was lying parallel to the bull, thus depriving him of his head-rest. He gave a low grunt of annoyance at this, and heaving up his bulk he flopped down on top of her back, so that she was half-hidden by his body. Then he closed his eyes and prepared for sleep. But his wife, with her spouse's great bulk half-covering her, had other ideas. She wriggled sideways so that the bull's barrel-shaped body slipped off her back and settled into the shingle with a scrunch. Then she leant forward and started to bite at his mouth and chin, very delicately, and in a slow and languorous manner. The bull kept his eyes tightly shut, and put up with these caresses, only occasionally snorting as if he were very embarrassed. But at last the female's love-play seduced him, and he opened his eyes and started to bite at the back of her glossy neck. With these signs of affection from her lord the female became as excited as a puppy, rolling and wriggling under his great head as he bit her, nibbling at his pigeon-chest and uttering subdued 'woofing' noises through her nose, so that her long whiskers stood out like fans of spun glass round her neat muzzle. As she writhed on the shingle he bent his head and delicately sniffed at her hindquarters, like some bloated old gourmet savouring the bouquet of a rare brandy. Then he hauled himself slowly and ponderously on top of her and entered her. Now she was straining up her face to his so that their whiskers entwined, biting his muzzle, his nose and his throat, and he in his turn engulfed her neck or her throat in savagely restrained bites. Their hindquarters undulated together, not quickly, urgently and crudely as in most animals, but slowly and carefully, the movements as smooth and precise as honey pouring from a jar. Presently, closely entwined, they reached their shuddering climax and then relaxed. The bull hauled himself off his wife and flopped down beside her, where they lay gently nibbling one another's mouths and faces with a tenderness that was remarkable. The whole act had been beautiful to watch, and was a lesson in restrained love-making which a lot of human beings would do well to emulate.

I have not as yet mentioned the fur seal pups which were such an important and amusing part of the colony. There were hundreds of them, and they moved continuously through the mass of sleeping, love-making, bickering adults, looking like animated black ink-blots. They would lie sleeping on the shingle in the most extraordinary abandoned attitudes, as though they were really balloon animals that had suddenly been half-deflated. Then, suddenly, one would wake up and discover that its mother was not there, and it would hoist itself on to its flippers and move sturdily down the beach, employing the strange rumba-like movement of the adult seal. Planting its flippers in the shingle with great determination, it would pause every few yards to open wide its pink mouth and bleat forlornly, like a lamb. Then, after it had wandered some distance in search of its parents, its bravado and strength would desert it, and it would give one more despairing bleat and then flop down on its tummy and sink almost immediately into a deep and refreshing sleep.

There appeared to be a rather vague crèche system in operation for some of the pups, for in places there would be groups of them, perhaps ten or twenty together, looking like heaps of curiously shaped coal. There would be a young bull or a couple of females sleeping nearby who were apparently in charge of these crèches, for if one of the babies wandered outside the invisible area that formed the crèche, one of the adults would rouse itself, undulate after it, catch it up in its vast mouth, give it a good shaking and throw it back into the nursery again. In spite of careful watching I was never able to decide satisfactorily whether these groups of babies were the progeny from one family of seals, or whether they were a mixture from several families. If they came from several families then these groups of babies would be, in effect, a sort of nursery-school or kindergarten where the babies were dumped while the parents went down to the sea to swim or feed. I wanted to film the daily behaviour of the pups, but in order to do this one had to pick out one particular baby, and as they were all identical in size and colour this was difficult. Then, just when I had begun to despair, I saw a pup that was recognizable. He had obviously been born later than the others, for he was only half their size, but what he lacked in inches he more than made up for in determination and personality.

When I first noticed Oswald (as we christened him) he was busily engaged in stalking a long ribbon of glittering green seaweed that lay on the shingle, and which he was obviously under the impression was some sort of monstrous sea-serpent which was threatening the colony. He shambled towards it, bleary-eyed, and stopped a yard or so away to sniff. A slight wind twitched the end of the seaweed, and at this obviously threatening display

Oswald turned and lolloped off as fast as his flippers would carry him. He stopped a safe distance away and peered over his shoulder, but the wind had died now and the seaweed lay still. Carefully he approached it again, stopping some six feet away to sniff, his fat little body taut and trembling, ready to run should he see the slightest movement. But the seaweed lay quiet in the sun, shining like a ribbon of jade. He approached it slowly and carefully, giving the impression that he was almost tiptoeing on his great flat flippers, and holding his breath in case of accidents. Still the seaweed made no movement. Cheered by this display of cowardice, Oswald decided that it was his duty to save the colony from this obviously dangerous enemy, which was liable to take them unawares. He shuffled his bottom to and fro ridiculously, so that his hind flippers got a good grip in the shingle, and then launched himself at the seaweed. In his enthusiasm he rather overshot the mark, and ended up on his nose in a fountain of shingle, but with a large section of the seaweed firmly grasped in his mouth. He sat up, the seaweed dangling from either side of his mouth like a green moustache, looking very pleased that his first bite had apparently disabled the enemy completely. He shook his head from side to side, making the weed flap to and fro, and then, shambling to his flippers, he galloped off along the beach trailing the weed on each side of him, occasionally shaking his head vigorously, as if to make sure his victim was really dead. For a quarter of an hour he played with the weed, until there was nothing left but a few tattered remnants. Then he flung himself down on the shingle, exhausted, the remains of the weed wound round his tummy like a cummerbund, and sank into a deep sleep.

Presently, when he woke up, he remembered that originally he had been looking for his mother, before his attention was distracted by the weed. So he shambled to his feet and made off down the beach, bleating soulfully. Suddenly in the middle of his grief he noticed a seagull squatting on the shingle near him. Forgetting about his mother he decided that the seagull should be taught a lesson, so he humped himself up indignantly and rumbaed towards it ferociously. The gull watched his approach from the corner of one cold, inimical eye. Oswald undulated across the shingle, panting a little, a look of grim determination on his face, while the gull watched him sardonically. Each time Oswald charged it side-stepped neatly, pattering a few paces on its webbed feet, with the air of a professional matador eluding a very inexperienced bull. Four times this happened, and then the gull grew bored. At the next charge he opened his wings, gave a couple of lazy flaps, and glided off down the beach to a more restful spot.

Oswald, the object of his wrath having vanished, suddenly remembered his mother and started out to search for her, bleating loudly. He made his way towards the most crowded part of the colony, a jumbled mass of cows and bulls all enjoying a siesta. Oswald ploughed his way through them, treading with complete impartiality on cows and bulls alike, scrambling over their backs, treading on their tails, and planting his flippers in their eyes. He left behind him a wake of infuriated adults who had been woken from a refreshing sleep by a large flipper covered with shingle being planted in the most vulnerable portion of their anatomy. At one point he discovered a cow lying on her back, exposing her teats to the rays of the sun, and he decided that it would be a suitable opportunity to stop for a snack. He had just taken a firm hold of one of the teats, and was preparing to imbibe life-giving nourishment, when the cow woke up and looked down at him. For a second she gazed at him fondly, for she was still half-asleep, but then she suddenly realized that he was not her son, but some dastardly interloper helping himself to a free drink. With a grunt of wrath she bent down, pushed her nose under his fat tummy, and, with a quick flip of her head, sent Oswald somersaulting through the air to land on the head of a sleeping bull. The bull was not amused, and Oswald had to be pretty nifty on his flippers to escape punishment. He plodded on over the mountain ranges of sleeping seals with grim determination. Then, at last, he slipped while negotiating a particularly rotund female, and fell on top of a young bull who was sleeping next door to her. The bull sat up, snorted indignantly, and then bent down and seized Oswald in his great mouth before the pup could get away. Oswald dangled there by the scruff of his neck, without movement, while the bull decided what was the best thing to be done with him. At last he decided that a little swimming lesson would do Oswald no harm, and so he flopped his way down to the sea, Oswald dangling from his mouth as limp as a glove.

I had often watched the bulls giving the pups swimming lessons, and it was a frightening sight. I felt quite sorry for Oswald. The bull paused at the edge of the surf and started to shake Oswald to and fro, until one felt certain that the pup's neck was broken, and then hurled him some twenty feet out into the waves. After a prolonged submersion Oswald surfaced, flapping his flippers desperately, spluttering and coughing, and struck out towards the shore. But the bull lumbered into the water and caught him by the neck again, long before he was in his depth, and then proceeded to hold him under the water for five or ten seconds at a time, eventually releasing his hold so that Oswald popped up like a cork, gasping for breath. After this had happened three or four times Oswald was so frightened and exhausted that he tried to attack the bull's great bulk with open mouth, uttering spluttering yarring cries. This, of course, had about as much effect as a pekinese attacking an elephant. The bull simply picked Oswald up, shook him well and flung him out to sea again, and repeated the whole process. Eventually, when it was obvious that Oswald was so exhausted that he could hardly swim, the bull took him into the shallows and let him rest for a little while, but standing guard over him so that he could not escape. When he was rested Oswald was picked up and thrown out to sea again, and the whole lesson was repeated.

This went on for half an hour and would have gone on longer, but another bull came and picked a quarrel with Oswald's instructor, and while they were fighting it out in the shallows

Oswald made his escape, scrambling back to shore as fast as he could, wet, bedraggled and thoroughly chastened.

These swimming lessons, as I say, were to be seen very frequently, and were agony to watch, for not only was the terror of the pups so piteous, but I was always convinced that the bulls might go too far and actually drown one of them. But the babies appeared to have the elasticity of mind and body that allowed them to survive these savage swimming lessons, and none of them seemed any the worse.

The adults spent ninety per cent of the day sleeping, and only occasionally the young bulls and cows would venture into the water, but it was not until evening that the colony as a whole went swimming. As the sun sank lower and lower a restlessness would prevail throughout the colony, and presently the females would hump themselves down to the water's edge, and the water ballet would begin. First two or three cows would enter the shallows and start swimming up and down, slowly and methodically. For some time the bull would watch them in a lordly manner, and then he would lift his huge bulk and shoulder his way into the surf with the air of a heavyweight boxer entering the ring. There he would pause and survey the sinuous shapes of his wives before him, while the foam made an Elizabethan ruff of white round his fat neck. His wives, desperately trying to get him to join in their game, would tumble and curve in the water ahead, their coats now gleaming and black with sea-water. Then, suddenly, the bull would submerge, his portly form disappearing beneath the water with a speed and grace that was startling. His blunt, snub-nosed head would appear between the bodies of his wives, and the entire picture would change. Whereas before the female's movements had been slow, gentle curvings of the body on the surface and beneath the water, now the tempo of their play quickened, and they would close in round the bull, making him the focal point of their game. Their movements as smooth as a flow of oil, they would curve over and under him, so that he was like a stocky maypole with the slim, swift ribbon of female seals drifting and fluttering around him. He would sit there with his massive head and neck out of the water, peering with supreme smugness into the sky, while his wives formed a whirlpool around him, weaving and gliding faster and faster, demanding his attention. Suddenly he would yield and, bending his head, he would open his mouth and bite playfully at a passing body. This was the signal for the ballet proper to begin.

The females' arrow-swift bodies and the bulk of the male would entwine like a gleaming black plait, curving and twisting through the water, assuming the most graceful and complicated shapes like a pennant whipped by the wind. As they rolled and curved through the water, leaving a foam-smudged track behind them, you could see them biting at each other with a sort of languorous lovingness, the gentle bites of affection, possession and submission. The tide would be coming in so gently that there was hardly any movement of the sea, but the seals would create in miniature their own seascape: sometimes they would slide free of the water, leaving no ripple on the surface, and at other times they would burst from the depths in a white rose of foam, their shining bodies curving up into the air like black boomerangs, before turning and plunging into the water again with a clean cut that scarcely disturbed the even surface.

Occasionally one of the young, unattached bulls would attempt to join one of these family groups in their play, and immediately the old bull would forget his game. He would submerge and suddenly reappear at the young bull's side in a crumble of foam, uttering a sort of gargling roar that had started beneath the surface. If the young bull was quick he would hurl himself sideways in the water, and the old bull's leap would be abortive and he would land on the water surface with a crack like a cannon going off, and the noise would roll and echo down the coast. Then it would be a question of who recovered first, the young bull from his awkward sideways leap, or the old bull from his belly-splitting charge. If the old bull recovered first he would seize the younger one by the neck and they would roll and thrash in the water, roaring and biting in a tidal wave of foam, while the females glided round them watching lovingly the progress of the battle. Eventually the young bull would break free from the savage grip of his adversary and plunge beneath the waves with the old bull in hot pursuit. But in swimming under water the young bull would have the slight advantage that he was not so bulky and therefore slightly faster, and he would generally escape. The old bull would swim pompously back to his wives and squat in the water, staring grandly up into the sky while they swam round him, reaching their pointed faces out of the water to kiss him, gazing at him with their huge melting eyes in an ecstasy of admiration and love.

By this time the sun would have sunk into a sunset of pink, green and gold, and we would make our way back to camp to crouch shivering over the fire, while in the distance, carried by the night wind, steady and bitterly cold, we could hear the noises of the seals, belching and roaring and splashing in the black and icy waters along the empty coast.

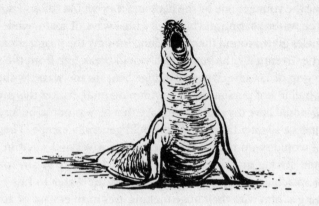

4.

The Bulbous Beasts

They did not remain long under water, but rising, followed us
with outstretched necks, expressing great wonder and curiosity.

CHARLES DARWIN: THE VOYAGE OF H.M.S. BEAGLE

After we had spent some ten days filming the fur seals I decided
that, reluctant though I was to leave these beautiful and fascinating animals, we really ought to move on and try and locate the
elephant seals before they left the peninsula in their southward
migration. So, for the next four days, we drove to and fro about
the peninsula searching for the *elefantería*, and seeing a variety of
wild-life, but no elephant seals.

I was amazed and delighted at the numbers of creatures we
saw on the Valdes peninsula. When I thought that, a few miles
away across the isthmus, lay hundreds of miles of scrub-land
which we had driven through without seeing a single living creature, and yet on the peninsula life abounded, it seemed incredible.

It was almost as if the peninsula and its narrow isthmus was a *cul-de-sac* into which all the wild-life of Chubut had drained and from which it could not escape. I wish that it were possible for the Argentine Government to make the whole peninsula into a sanctuary, for which it seems to have been designed by nature. To begin with you have a wonderful cross-section of the Patagonian fauna, all concentrated in a limited area, and most of it very easy to see. Secondly the whole area could be easily and effectively controlled by virtue of the narrow isthmus connecting it to the mainland; a check point on this could keep an adequate control on the people who entered and left the area, and keep an eye out for the sort of 'sportsmen' (of which there are some in every country throughout the world) who would think it fun to chase guanaco in fast cars, or pepper the bull fur seals with buckshot. I do not think that the fact of the peninsula being divided up into several large sheep *estancias* is of great importance. True, the guanaco and the fox are shot, the first because it is supposed to eat grazing that would be better employed feeding sheep, and the latter because it is big enough to take lambs and chickens. Yet, in spite of having the *estancias* against them, both these animals, when we were there, seemed very common. Provided the sheep-farmers behaved sensibly, I think a balance between the domestic and the wild animals could be maintained. If the peninsula could be declared a wild-life sanctuary now, then, when Southern Argentina is opened up still further (which seems inevitable), and when decent roads make the peninsula less inaccessible, it could well turn out to be a tourist attraction of considerable value.

In our search for the *elefantería* we covered a lot of the peninsula, and the commonest bird we saw was undoubtedly the martineta, a species of tinamu. It is plump, a partridge-shaped little bird, about the size of a bantam. Its plumage is a rich array of autumn browns, speckled and streaked with golds, yellows and creams in an intricate and lovely pattern. Its cheeks are a pale cream colour, with two black stripes showing up well on this background, one running from the corner of the eye to the neck and the other running from the edge of the beak to the neck. On its head there is an elongated crest of dark feathers, which curves like a half-moon over its head. It has large, dark eyes, and a general air of innocent hysteria.

Martinetas were to be seen everywhere along the rough roads in little groups of five or ten. Ridiculously tame, they would stand in the middle of the road, watching the Land-Rover's approach with wide eyes, bobbing their heads so that their silly crests twitched and fluttered, not bothering to move until you slowed down within a few feet of them and blew the horn. Then, stretching out their necks and holding their heads low, as if searching the ground for something they had lost, they would scuttle off into the scrub. They were most reluctant to fly, and in order to make them do so you had to pursue them for considerable distances through the undergrowth. Then, when they felt you were coming too near, they would launch themselves into the sky with an air of desperation. It was a curious, laboured flight, like that of a bird which has never learnt to use its wings properly. They would give four or five frantic flaps of their wings, and then glide until their fat bodies had almost dragged them to earth again, when they would give another series of wild flaps and then glide on a bit further. As they flew the rush of wind through their feathers produced a curious wailing note, that rose and fell flute-like, as they flapped and glided away. Their partiality for sitting in the middle of the road was due to the fact, I think, that it was only on these bare earth surfaces that they could construct the best dust-baths. In many places they had scooped out quite deep depressions in the red earth, and you could see three or four of them standing patiently waiting their turn, while one member of the flock rolled and kicked absurdly in the bath, fluttering its wings to throw the dust over its body.

These lovely, slightly imbecile birds are, of course, ground-nesting, and I think that they themselves, their eggs and their young, form an important item of diet among the carnivorous mammals of the peninsula, particularly the pampas fox, which was a common predator in the area. They are slim, grey, dainty little animals, with incredibly slender and fragile-looking legs. They appeared to hunt as much by day as by night, and were usually to be seen in pairs. They would suddenly dash across the road in front of us as we drove along, their bushy tails streaming out behind them like puffs of grey smoke, and on reaching the other side of the road they would skid to a halt and, squatting on their haunches, examine us craftily.

At one of the places in which we camped a pair of these little foxes paid us a visit, the only animal apart from the guanaco to do so. It was about five in the morning, and from my bed under the rear axle of the Land-Rover I was watching the sky turn green with dawn, while, as usual, trying to pluck up the courage to quit the warmth of my blankets and light the fire for breakfast. Suddenly, from the yellow scrub around us, the two foxes appeared as unexpectedly and as silently as ghosts. They approached the camp cautiously, with the conspiratorial air of a couple of schoolboys raiding an orchard, with many pauses to sniff the dawn wind. It was fortunate, at that precise juncture, that no one was snoring. I can put it on record that there is nothing quite so effective for scaring off wild animals as three women in the back of a Land-Rover, all snoring in different keys.

Having circled the camp without mishap, they grew bolder. They approached the ashes of the fire, sniffed at them deeply, and then frightened each other by sneezing violently. Recovering from this shock they continued their investigation and found an empty sardine tin, which, after a certain amount of low bickering, they proceeded to lick clean. Their next discovery was a large roll of bright pink toilet paper, one of the few luxury articles in our equipment. Having proved that it was not edible, they then discovered that if it was patted briskly with a paw it unravelled itself in the most satisfactory manner. So, for the next ten minutes,

they danced and whirled on their slender legs, hurling the toilet roll to and fro, occasionally taking streamers of it in their mouths and leaping daintily into the air, returning to earth with the paper wrapped intricately round their necks and legs. This game was conducted so silently and so gracefully that it was a delight to watch, and their agile bodies were well set off against the green sky, the yellow-flowered bushes and the pink paper. The whole camp site was taking on a gay carnival air, when somebody in the Land-Rover yawned. The foxes froze instantly, one of them with a piece of toilet paper dangling from his mouth. The yawn was repeated, and the foxes vanished as silently as they had come, leaving – as a souvenir of their visit – some hundred and twenty feet of pink paper fluttering in the breeze.

Another creature that we saw very frequently was the Darwin's rhea, the South American counterpart of the African ostrich. These birds were smaller than the rheas from Northern Argentina, more delicate in build and a more pearly grey in colour. They were generally in small flocks of five or six, and on many occasions we saw them moving through the scrub in conjunction

with a flock of guanaco. I think one of the loveliest sights we saw on the peninsula was a herd of six guanaco with three graceful, cinnamon-coloured babies, trotting slowly through the golden scrub in company with four Darwin's rheas, who were ushering along a swarm of twelve young, each dressed in its striped baby plumage, so that they looked like a line of tiny fat wasps running close to their parents' great feet. While the baby rheas were very sedate and orderly, like a school crocodile, the baby guanacos were more exuberant and unruly, dancing about in amongst the adults, in exciting, daring and complicated gambols. One of them carried out such an intricate gambol that he bumped into one of the adults and received a sharp kick in the stomach as punishment, after which he became very subdued and trotted quietly along behind his mother.

If undisturbed the rheas would pace along in a very regal manner. But, occasionally, we would come upon them when they were on the road and immediately panic would ensue. Instead of swerving off into the scrub, they would set off in a disorderly cluster down the road, running with the slightly effeminate grace of professional footballers. As we drove the Land-Rover closer and closer they would increase their speed, lowering their long necks groundwards, their feet coming up so high with each step that they almost touched what passes for a chin in a rhea. One I paced in this manner ran six feet in front of the Land-Rover bonnet for a distance of half a mile, averaging between twenty-five and thirty miles an hour. Eventually, when you had followed them like this for some considerable time, it would suddenly occur to them that they might be safer in the scrub. So they would put on a sudden burst of speed, open their pale wings in a graceful gesture, swerve off the road with a ballet-like grace and go bouncing away into the distance.

These rheas, like the common rhea of the north, have communal nests, that is to say several females lay their eggs in one nest. This is a mere scrape in the ground, lined with some dry grass or a few twigs, and you can find as many as fifty eggs in the one nest. As in the common rheas the male Darwin's rhea does the hard work of incubating the eggs and rearing the young when they hatch. The highly-polished eggs are a fine green colour when just laid, but the side that is towards the sun soon fades, first to a dull mottled green, then yellowish, then to pale blue and finally to white. The rheas are so prolific that their eggs, and, to a large extent, their young, form an important item of diet for the predators of the peninsula.

Another creature which was very common, and which we frequently met on the roads, was the *pinche* or hairy armadillo. We saw them just as much by day as by night, but the time they were most frequently seen was towards evening in the rays of the setting sun, trotting to and fro over the road surfaces, sniffing vigorously, looking like strange clockwork toys, for their little legs moved so fast they were a mere blur beneath the shell. They are fairly thickly cloaked with long, coarse white hair, but I should not have thought that this would have provided them with any protection from the cold in the winter. I presume they must

hibernate in the winter months, for there could be nothing for them to eat as the ground is frozen to a depth of several feet. All the ones we caught were covered with a tremendously thick layer of fat, and their pale pink, heavily wrinkled bellies were always bulging with food. Their main diet must consist of beetles, their larvae, and the young and eggs of ground-nesting birds like the martineta, though sometimes they may come across a windfall in the shape of a dead sheep or guanaco. Frequently they could be seen right down on the sea-shore, trotting briskly along the tide line, looking like small, rotund colonels on a Bournemouth sea-front, imbibing the health-giving ozone, though they would occasionally spoil the illusion by stopping to have a light snack off a dead crab, a thing I have never seen a colonel do.

Watching all this wild-life was, of course, fascinating, but it was still not bringing us any nearer to our objective, which was the elephant seals. We had, by now, covered quite a large area of the coast, without any success, and I began to think we were too late, and that the elephant seals were already drifting southwards towards Tierra del Fuego and the Falkland Islands. But just when I had given up hope we discovered an *elefantería* which no one had told us about, and then we only found it by luck. We had been walking along a fairly high cliff, pausing every quarter of a mile or so to examine the beach below us for signs of life. Presently, we rounded a small headland and came to a bay where the beach at the base of the cliff was covered in a tumbling mass of rocks. Some of these rocks were so large that, from our vantage point, we could not tell what might be lying behind them, so, after searching along the cliff for a short way, we found a rough path which led us down to the shore, and made our way down to investigate.

The beach was of bright mottled shingle, each pebble sea-polished so that it shone in the evening sun. The boulders, some as large as a cottage, lay tumbled haphazardly along the beach, grey and fawn in colour. Some of them were so large and fretted into such weird shapes by the wind and the sea, that it was a major operation scrambling over them, weighted down as we were with the cameras and equipment. We struggled through and over them for some distance, and then decided that what we needed was food. So, choosing a rock that had been moulded to make a natural seat, we sat down and unpacked our food and wine. I was convinced by then that there was not an elephant seal for miles, and I was thoroughly depressed and irritated with myself for having spent so much time on the fur seals.

'Well, we might find some tomorrow,' said Jacquie soothingly, handing me a sandwich that appeared to have three-quarters of the Patagonian topsoil adhering to it.

'No,' I said, viewing this sustenance with a jaundiced eye, and refusing to be comforted, 'they've gone south now. They've had their babies and left. If I hadn't spent so much time on those damned fur seals we might have found them.'

'Well, it's your own fault,' said Jacquie logically. 'I kept telling you that you had enough film of the fur seal, but you kept insisting we spent just one more day.'

'I know,' I said gloomily, 'but they were such wonderful creatures, I couldn't tear myself away.'

Marie, with the air of one who is making the best of a disaster, seized a bottle of wine, and as the cork popped out of the bottle a large, slightly elongated and egg-shaped boulder some ten feet

away gave a deep and lugubrious sigh, and opened a pair of huge, gentle, liquid-looking eyes of the deepest black, and gazed at us placidly.

Once it had thus revealed itself as an elephant seal, one wondered why one had ever thought it was anything else; and a close and excited scrutiny of the surrounding beach showed us that we were, in fact, sitting next to twelve of the gigantic beasts, which had all remained calmly sleeping while we had walked up to them, seated ourselves, and unpacked our food like trippers at Margate. They so closely resembled the rocks amongst which they lay that I began to wonder how many other groups we had walked past in our search for them. After watching the fur seals, I had expected the elephant seal colony to be a much more boisterous and vivacious lot, whereas here they were, lying about the beach in attitudes of relaxed abandon, displaying about as much boisterousness as could be expected from a convention of dropsy sufferers having a chess tournament in a Turkish bath. We walked among the huge, snoring carcases, and by investigation we discovered that of the twelve animals there, three were males, six were females, and three were well-grown young. The babies measured about six feet in length, and the females about twelve to fourteen feet. The real bulk was reserved for the males. Two of these were young bulls, each about eighteen feet in length, while the last was a fully adult bull, and measured twenty-one feet in length.

This bull was a magnificent beast, with a huge barrel-like body, and a great carunculated nose, like that of a confirmed gin-drinker. He lay on the shining shingle like a colossal blob of putty, occasionally sighing deeply so that his nose wobbled like a jelly, or every so often waking up sufficiently to ladle some damp shingle on to his back with one of his flippers. His placidity towards our intrusion was extraordinary, for we approached within three or four feet to measure and take photographs, and all he did was to open his eyes, survey us dreamily, and sink back into sleep again.

For me this was a tremendously exciting experience. Other people may have a burning ambition to see the Leaning Tower of Pisa, or visit Venice, or see the Acropolis before they die. But my ambition had been to see a live elephant seal in his natural environment, and here I was, lying on the shingle eating sandwiches within five feet of one, who lay there looking not unlike a baby barrage balloon which has, unaccountably, been filled with dough. With a sandwich in one hand and a stop-watch in the other I checked on his breathing, which is one of the many remarkable things about an elephant seal. They breathe fairly regularly some thirty times during five minutes, and then they stop breathing for a time, which varies from five to eight minutes. Presumably this is of great use to them when they are at sea, for they can rise to the surface, breathe, and then sink below the water and hold their breath for this considerable period without having to resurface and refill their lungs. I was so carried away, lying there with this gigantic and fantastic animal within touching distance, that I proceeded to give the others a lecture on the elephant seal.

'It's quite extraordinary the soundness of their sleep. Do you know there was one naturalist who actually went and lay on top of an elephant seal without waking it?'

Jacquie surveyed the colossal animal in front of me.

'Rather him than me,' she said.

'Apparently the females don't become sexually mature until they are two years old. They have this delayed implantation thing as well . . . you know, where they're mated and retain the sperm for varying lengths of time in their bodies before allowing it to develop. Now those babies over there are this year's brood. That means they won't be ready to breed . . .'

'*This* year's brood?' Jacquie interrupted in astonishment. 'I thought they were about a year old.'

'No, I should say they are four or five months old.'

'How big are they when they're born, then?'

'Oh, about half that size, I should think.'

'Good God!' said Jacquie with feeling. 'Fancy giving birth to a thing *that* size.'

'There you are,' I said. 'It just goes to show that there's always someone worse off than you are.'

The elephant seal, as if in agreement, gave a deep, heart-rending sigh.

'Do you know that the intestine of an adult bull can measure six hundred and sixty-two feet?' I inquired.

'No, I didn't,' said Jacquie, 'and I think we'd all enjoy our sandwiches more if you refrained from divulging any more secrets of their internal anatomy.'

'Well, I thought it would interest you.'

'It does,' said Jacquie, 'but not when I'm eating. It's the sort of information I prefer to acquire between meals.'

There were several things that struck one immediately about the elephant seals, once one had got over incredulity at their mere size. The first thing was, of course, their ridiculous hind-quarters. The fur seal (which is really a sea lion) has the hind limbs well developed as legs, so that when they move they hoist themselves up on to all four legs and walk as a dog or a cat would. But in the elephant seal, which is a true seal, the hind limbs are minute and pretty useless, with stupid flippers that make it look as though the animal has had a couple of empty gloves attached to its rear end. When the creature moves all the propulsion comes from the front flippers, and the humping of the massive back, a slow, ungainly method of movement that was painful to watch.

There was quite a colour variation among the herd. The old bull was a rich, deep slate-grey, tastefully speckled here and there with green, where some marine algae was apparently growing on his tough hide. The young bulls and the cows were a much paler grey. The babies were not bald and leathery like their parents, but each was wearing a fine fur coat of moon-white hair, close and tight as plush. The adults had so many folds and wrinkles all over them that they looked rather as if they were in need of a square meal to fill out the creases, as it were, whereas the babies were so rotund and glossy they looked as though they had all just been blown up with bicycle pumps, and would, if they were not careful, take to the air.

From the point of view of filming, the elephant seal colony was, to say the least, difficult. All they wanted to do was sleep. The only real movement they made was to open and close their huge nostrils as they breathed, and occasionally one would shovel some shingle on to its back; but as there was no preliminary warning to this action it took me some time to get it on film. Sometimes one of them would hump itself forward, eyes tightly shut, burrowing its great nose through the shingle like a bull-dozer. Even when I had got all these actions on film it still did not seem to me that the elephant seals were showing themselves to advantage; they lacked action, which, after all, is one of the things necessary for a moving picture. One of the extraordinary things about these seals is the flexibility of the backbone. In spite of their bulk and vast quantities of blubber, they can bend themselves backwards, like a hoop, until the head touches the uplifted tail. How to get them to demonstrate this for me to film, when they were all lying about displaying the animation of a group of opium smokers, was somewhat of a puzzle. At last, however, we were successful with the old bull, by the simple expedient of throwing handfuls of fine gravel on to his tail. The first handful made him stir slightly and sigh deeply, without opening his eyes. The second handful made him open his eyes and stare at us in mild surprise. With the third handful he raised his head, drew back his snout so that it wrinkled like a concertina, opened his mouth and uttered a hissing roar, and then fell back on to the shingle as if exhausted by this effort and went back to sleep again.

Eventually, however, our bombardment got on his nerves. It did not, of course, hurt him, but a constant rain of shingle on your rear-end when you are trying to get to sleep can be extremely irritating. He suddenly became very wide awake and reared up so that he was like the letter J with his head high in the air, his mouth opened wide uttering the loud hissing roar, an oddly reptilian sound for such a monstrous mammal to make. Four times he reared up like this, and then, seeing that the display was having no detrimental effect on our morale, he did what all seals do in moments of crisis: he burst into tears. Great, black tears oozed out of his eyes and trickled forlornly down his cheeks. He lowered himself full length on to the shingle, and proceeded to move backwards towards the sea, like a gargantuan caterpillar, humping his body up with tremendous effort, the fat along his back rippling into waves as he moved. At last, with a final plaintive roar and another flood of tears, he backed into the water, and an incoming wave broke in a garland of white foam around his shoulders. The rest of the herd became alarmed at their lord and master's disappearance, and they all raised their heads and started to look at us uneasily. Then one of the babies panicked, and hunched its way down to the sea, tears streaming down its white face. This was the final straw, and within a minute the whole herd was rushing seawards, looking like a flock of huge maggots in pursuit of a cheese.

Sadly we packed up our equipment and started up the cliff, sadly because we had just completed our last task, and this meant that we must leave the peninsula with its wonderful animal life, and head back to Buenos Aires and the next stage of the expedition. As we made our way along the twilit cliff path we saw the old bull elephant seal for the last time. His head appeared out of a wave, his dark eyes surveyed us puzzledly. He snorted, a reverberating noise that echoed along the cliffs and made his nose vibrate. Then, still watching us sadly, he sank slowly beneath the icy waters and disappeared.

PART TWO

The Customs of the Country

The plane taxied out across the darkened airfield to where the runway lay, between two strips of diamond-bright lights. Here it paused, revved up its engine until every bone in the plane's metal body seemed to screech out in protest, and then suddenly rushed forward. The strip-lights fled past, and then suddenly we were airborne, the plane tipping from side to side like a slightly drunken swallow as it climbed higher and higher. Then, below me, Buenos Aires lay spread in the warm night like a chessboard of multicoloured stars. I unfastened my safety belt, lit a cigarette and lay back in my seat, feeling very mellow and full of farewell brandy. At last I was on my way to a place I had long wanted to visit, a place with a magical name: Jujuy.

When we had returned from the south the effects of the car crash we had had soon after arrival in Argentina (in which Jacquie was the only one hurt) had begun to make themselves felt; the terrible jolting we had undergone on the Patagonian roads, and

91

the rough conditions under which we had been forced to live, had resulted in her getting blinding headaches. It was obvious that she could not continue the trip, so we had decided to send her back to England. She had departed the week before, and this left Sophie and me to finish the trip. So, while Sophie remained in our little villa with its garden already stuffed with animals which she had to minister unto, I made tracks for Jujuy, to try and add to the collection.

As the plane droned on through the night I dozed in my seat and tried to remember all I knew about Jujuy, which was precious little. It is a north-western province of Argentina, bordered on the one side by Bolivia, and on the other by Chile. It is a curious place in many ways, but chiefly because it is like a tropical tongue, as it were, inserted into Argentina. On the one side you have the mountains of Bolivia, on the other the curious, desiccated province of Salta, and between the two the lush tropical area of Jujuy, which compares favourably with anything to be found in Paraguay or southern Brazil. Here I knew that you could find the colourful, exciting tropical fauna, just starting to encroach on the Pampa and grassland fauna, and it was these creatures I was after. Thinking about all these magnificent animals I fell into a deep sleep, and was just dreaming that I was catching a particularly malevolent jaguar with a lassoo, when I was awoken by the steward shaking my arm. Apparently we had arrived at some godforsaken place, and all passengers had to dismount while the plane refuelled. Plane travel has never been my favourite form of transportation (except for very small planes, where you get a real sense of flying), so to be roused from a brandy-soothed sleep at two in the morning and be forced to stand around in a tiny bar that did not offer anything more exciting than lukewarm coffee did not improve my temper. As soon as they would allow I got on the plane again, settled down in my seat and tried to sleep.

Almost immediately I was roused by what appeared to be a ten-ton weight, descending on my arm. I extricated it with difficulty, before any bones were broken, and glared at the person responsible. This was not very effective, as the interior of the plane was lit by what appeared to be a series of fireflies suffering from pernicious anaemia. All I could see was that the next seat to me (until then mercifully empty) was now being inundated – there is no other word – by a female of colossal proportions. The various portions of her anatomy which she could not cram into her own seat she had generously allowed to overflow into mine.

'*Buenas noches,*' she said pleasantly, exuding sweat and scent in equal quantities.

'*Buenas noches,*' I mumbled, and hastily closed my eyes and huddled into what was left of my seat, in order to put an end to the conversation. Fortunately, my companion, after this exchange of pleasantries, settled herself down for sleep, with much grunting and shifting and deep shuddering sighs that were vaguely reminiscent of the elephant seals. Presently she twitched and mumbled her way into sleep, and then started a prolonged and interesting snore that sounded like someone rhythmically rolling small potatoes down a corrugated iron roof. Lulled, rather than disturbed by this sound, I managed to drop off myself.

When I awoke it was light, and I surreptitiously examined my still sleeping companion. She was, as I say, a fine figure of a woman – all twenty stone of her. She had clad her generous body in a silk dress in yellow and green, and she was wearing scarlet shoes, both now reclining some distance from her feet. Her hair was bright glossy black and carefully arranged in tiny curls all over her head, and to crown this she was wearing a straw hat to which half the fruit and vegetable produce of Argentina appeared to have been attached. This breath-taking horticultural achievement had slipped during the night, and now reclined over one of her eyes at a saucy angle. Her face was round and dimpled, and separated from her ample bosom by a lava-flow of chins. Her hands, I noticed, were folded demurely in her lap, and though they were reddened and work-roughened they were tiny and beautifully formed, like the hands of so many fat people. As I was watching her she suddenly gave a great, shuddering sigh and opened large, pansy-dark eyes and gazed about her with the vacant expression of an awakening baby. Then she focused on me and her dumpling face spread into a dimpled smile.

'*Buenos dias, señor*,' she said, inclining her head.

'*Buenos dias, señora*,' I replied, also inclining my head gravely.

From somewhere under the seat she hoicked out a handbag the size of a small cabin-trunk and proceeded to repair the damage that the night's sleep had done to her face. This was little enough, as far as I could see, for her complexion was as perfect as a magnolia petal. Satisfied at last that she was not going to let her sex down, she put away her bag, resettled her bulk, and turned her bright, kindly eyes on me. Wedged as I was there was no escape.

'Where are you travelling to, señor?' she asked.

'Jujuy, señora,' I replied.

'Ah, Jujuy?' she said, opening wide her dark eyes and raising her eyebrows, as though Jujuy was the most interesting and desirable place in the world.

'You are German?' she asked.

'No, English.'

'Ah, English?' with again the delighted surprise, as though to be English was something really special.

I felt it was time I took a more active part in the conversation. 'I do not speak Spanish at all,' I explained, 'only a very little.'

'But you speak *beautifully*,' she said, patting my knee, and then qualified it by adding, 'and I will speak slowly so that you may understand.'

I sighed and gave myself up to my fate; short of jumping out of the window on my left there was nothing else I could do. Having decided that my knowledge of Spanish was limited she came to the conclusion that I would get a better grasp of her conversation if she shouted, so now the whole plane was party to our exchange of confidences. Her name, it appeared, was Rosa Lillipampila and she was on her way to visit her married son in Salta. She had not seen him for three years, and this was to be a terrific reunion. This was also her first flight in a plane, and she was taking a child-like delight in it. She kept breaking off her conversation with shrill cries (which made the more nervous of the other passengers jump) in order to lean over me, enveloping me in scent and bosom, to peer at some landmark passing below. Several times I offered to change seats with her, but she would not hear of it. When the steward came round with morning coffee she fumbled for her bag to pay, and when it was explained that it was free she was so delighted that you would have thought the rather grubby paper cup full of gritty liquid was a magnum of champagne which the benevolent air company had bestowed upon her. Presently the red lights went on to tell us that we were landing yet again at some obscure township to refuel, and I helped her struggle to get the safety belt round her enormous girth. This was a strenuous task, and her shrieks of merriment at our efforts echoed up and down the plane.

'You see,' she panted, between gusts of laughter, 'when one has had six children and one likes to eat, one loses control over the size of one's body.'

At last, just as the plane touched down, we got the belt hitched round her.

We clambered out on to the tarmac, stiff and crumpled, and I found that my girl-friend moved with the grace and lightness of a cloud. She had obviously decided that I was to be her conquest of the trip, and so, with a courteous, old world gesture, I offered her my arm, and she accepted it with a beaming, coquettish smile. Linked together like a courting couple we made our way towards the inevitable small café and toilets that decorated the airport. Here she patted my arm, told me she would not be long, and drifted to the door marked 'Señoras', through which she passed with difficulty.

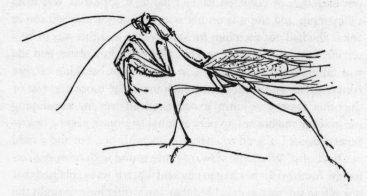

While she was communing with nature I took the opportunity
to examine a large bush which grew alongside the little café. It
was about the size of the average hydrangea, and yet on its
branches and among its leaves (after only a cursory inspection) I
found fifteen different species of insect and five species of spider.
It was obvious we were nearing the tropical area. Then I spotted
a very old friend of mine, a praying mantis, perched on a leaf,
swaying from side to side and glaring about with its pale, evil
eyes. I detached it from its perch and was letting it stalk its way up
my arm, when my girl-friend returned. On seeing the creature
she let out a cry that could, with a following wind, have been
heard in Buenos Aires, but, to my surprise, it was not a cry of
horror, but a cry of delighted recognition.

'Ah, the Devil's Horse!' she cried excitedly. 'When I was a child
we often used to play with them.'

This interested me, for, as a child in Greece, I used to play with
them as well, and the local people had also called them the Devil's
Horse. So, for ten minutes or so, we played with the insect, making it run up and down each other's arms, and laughing immod-
erately, so that all the other passengers obviously doubted our
sanity. At last we returned the mantis to his bush and went to
have a coffee, but just at that moment an official arrived, and with

much apologetic hand-spreading informed us that we would be delayed two hours. Groans of rage rose from the assembled passengers. There was, however, the official went on, a company bus which would run us into town, and there, at a hotel, the air company had arranged for us to have anything we wanted at their expense. My girl-friend was delighted. Such generosity! Such kindness! I helped her into the bus, and we rattled over the dusty road into the town and drew up outside a curiously Victorian-looking hotel.

Inside, the hotel was so ornate that my lady friend was quite overcome. There were huge, brown, imitation marble pillars, pots and pots of decayed-looking palms, flocks of waiters who looked like ambassadors on holiday, and a sort of mosaic of tiny tables stretching away, apparently, to the farthest horizon. She held very tight to my arm as I steered her to a table and we sat down. All this splendour seemed to bereave her of speech, so in my halting Spanish I ordered lavishly from one of the ambassadors (who did not appear to have shaved since his last official function) and settled back to enjoy it. Soon, under the influence of five large cups of coffee with cream, a plate of hot *medialunas* and butter, followed by six cream cakes and half a pound of grapes, my companion lost her awe of the place, and even ordered one of the ambassadors to fetch her another plate to put her grape-pips on.

Presently, replete with free food, we made our way outside to the coach. The driver was sitting on a mudguard, moodily picking his teeth with a matchstick. We inquired if we were now ready to return to the airport. He gazed at us with obvious distaste.

'*Media hora,*' he said, and returned to the cavity in his back molar, in which he obviously hoped to find a rich deposit of something, maybe uranium.

So my girl-friend and I went for a walk round the town to kill time. She was delighted to have this chance to act as guide to a real foreigner, and there was nothing she did not show and explain to me. This was a shoe-shop . . . see, there were shoes in the window, so one knew, without a shadow of a doubt, that it was a shoe-shop. This was a garden, in which they grew flowers. That was a donkey, over there, that animal tethered to a tree. Ah, and here we had a chemist's shop, where you purchased medicines when you were not well. Oblivious to the people trying to force their way past on the pavement, she insisted on standing in front of the chemist's window and giving such a realistic display of suffering that I expected someone to call for an ambulance, if the town boasted of such an amenity. Altogether our tour was a great success, and I was quite sorry when we had to return to the bus and be driven back to the airport.

Once more in the plane we had the Herculean task of lashing her into her seat, and then unlashing her once we were airborne on the last leg of our journey. Hitherto the country we had been flying over had been typical Pampa, with here and there an occasional outcrop of small hills, but by and large the view from the plane had been flat and featureless. But now the hills became more and more frequent, and higher and higher, covered with scrub and gigantic cacti like huge green surrealist candelabra. And then the air-pockets started.

The first was quite a big one, and one felt one's stomach had been left at least a hundred feet up as the plane dropped. My companion, who had been in the middle of an intricate and – to me – almost incomprehensible story about some remote cousin, opened her mouth wide and uttered a cry of such a piercing quality that the whole of the aircraft was thrown into confusion. Then, to my relief, she burst into peals of happy laughter.

'What was that?' she asked me.

I did my best, in my limited Spanish, to explain the mysteries of air-pockets, and managed to get the basic fact across to her. She lost all interest in the story about her cousin, and waited expectantly for the next air-pocket to make its appearance so that she could enjoy it to the full, for, as she explained, she had not been prepared for the first one. She was soon rewarded with a real beauty, and greeted it with a scream of delight and a flood of delighted laughter. She was like a child on a switchback in a fair, and she treated the whole thing as a special treat which the air company had provided for her enjoyment, like the meal we had just eaten. The rest of the passengers, I noticed, were not treating the air-pockets in the same light-hearted way, and they were all glowering at my fat friend with faces that were growing progressively greener. By now we were flying over higher and higher ground, and the plane dropped and rose like a lift out of control. The man across the gangway had reached a shade of green I would not have thought the human countenance could have achieved. My friend noticed this too, and was all commiseration. She leant across the gangway.

'Are you ill, señor?' she inquired. He nodded mutely.

'Ah, you poor thing,' she said and burrowing into her bag produced a huge bag of very sticky and pungent sweets which she thrust at him.

'These are very good for sickness,' she proclaimed. 'Take one.'

The poor man took one look at the terrible congealed mass in the paper bag and shook his head vigorously. My friend shrugged, gave him a glance of pity, and popped three of the sweets into her mouth. As she sucked vigorously and loudly she suddenly noticed something that had escaped her sharp eyes before, the brown paper bag in a little bracket attached to the back of the seat in front of us. She pulled it out and peered inside, obviously wondering if some other magnificent largesse from the kindly air company was concealed inside it. Then she turned a puzzled eye on me.

'What is this for?' she asked in a penetrating voice.

I explained the necessity of the paper bag. She held it aloft and examined it minutely.

'Well,' she said at last, 'if I wanted to get sick I should want something *much* larger than *that*.'

The man across the gangway cast a look at her ample form and the size of the brown paper bag, and the vision conjured up by her words was obviously too much for him, for he dived precipitously for his own bag and buried his face in it.

When the plane eventually touched down my girl-friend and I were the only ones who dismounted without looking as though we had just been through a hurricane. In the foyer of the airport her son was waiting, a pleasant-faced man who was identical in shape to his mother. Uttering shrill cries they undulated towards each other and embraced with a fat-quivering crash. When they surfaced, I was introduced and commended for the care I had taken of my protégée *en route*. Then, because the driver who was to meet me was nowhere to be seen, the entire Lillipampila family (son, wife, three children and grandmother) hunted round the airport like foxhounds until they found him. They saw me to the car, embraced me, told me to be sure to look them up when I was in Salta, and stood, a solid façade of fat, beaming and waving as I drove off on my way to Calilegua, the place where I was to stay. Kindness in Argentina is apt to be overwhelming, and after having been embraced by the entire Lillipampila family I felt every bone in my body aching. I gave the driver a cigarette, lit one myself and leant back and closed my eyes. I felt I deserved a few moments' relaxation.

5.
Jujuy

The elegance of the grasses, the novelty of the parasitical plants, the beauty of the flowers, the glossy green of the foliage, but above all the general luxuriance of the vegetation, filled me with admiration.

CHARLES DARWIN: THE VOYAGE OF H.M.S. BEAGLE

Calilegua was primarily a sugar-producing estate, though it also grew a certain amount of the more tropical varieties of fruit for the Buenos Aires market. It was a flat plain that lay cupped in a half-moon of mountains that were covered with thick, tropical forest. It was curious how suddenly you came upon this lushness of vegetation. We left the airport and for the first hour or so drove through a desiccated landscape of semi-eroded hills, sun-baked, scrub-covered, dotted here and there with the great swollen trunks of the *palo borracho* trees, their bark as thickly covered with spines as a hedgehog's back, and here and there you saw one of the giant cacti rearing up, perhaps twenty feet high, decorated with strange curving branches. These again were spine-covered and unfriendly. Then we sped round a couple of corners, down a hill and into the valley of Calilegua, and the vegetation changed, so suddenly that it was almost painful to the eye. Here were the vivid greens of the tropics, so many shades and some of such viridescence that they make the green of the English landscape look grey in comparison. Then, as if to assure me that I was back in the tropics, a small flock of parakeets swooped across the road, wheezing and chittering. Shortly afterwards we passed a group of Indians, dressed in tattered shirts and trousers and gigantic straw hats. They were short and squat, with Mongolian features and the curious sloe-coloured eyes over which there seems to be a bloom like a plum that covers thought and expression. They glanced at the car incuriously as we passed. After being among Europeans so long, and in the flat scenery of the Pampa and Patagonia, the Indians, the parakeets and the vividness of the country through which we drove went to my head like wine.

Presently the driver slowed down and swung off the main road on to a rough track that was thickly lined on both sides by thick clumps of gigantic bamboos, some of the canes being as thick as a man's thigh and pale honey-coloured, tiger-striped with green. These huge canes bent gracefully over the road and intertwined their fluttering green leaves overhead so thickly that the road was gloomy, and it was like driving down the nave of a cathedral. The sunlight flickered and flashed between the giant stems as we drove down the rutted track, and above the noise of the car engine I could hear the strange groans and squeaks that bamboos make when swayed by the wind. Presently we came to a villa half-hidden in a riot of flowers and creepers, and here the car stopped. Joan Lett, who, with her husband Charles, had invited me to Calilegua, came out to greet me, took me inside and gave me the most welcome cup of tea. Presently, when Charles returned from his work, we sat on the balcony in the fading indigo evening light and discussed my plan of campaign.

It has always been my experience in most parts of the world that if you go to an area which is fairly well populated you can obtain most of your common local fauna without much difficulty, for the local people keep the creatures either as pets, or rear them until they are old enough to form the basis for a meal. So your first job should be to go round every ranch and village in the vicinity and buy what you can. Then you can review your collection and try and fill in the gaps (which are generally the rare creatures) yourself. I propounded this philosophy to Charles, as the ice tinkled musically in our gin-and-tonics, but to my dismay he was not inclined to agree with me. He said that he did not think the Indians in Calilegua kept anything in the livestock line, except the usual run of cats, dogs and chickens. However, he promised that the next day he would get one of his more intelligent helpers to make inquiries in the village, and let me know the result. I went to bed fortified by gin but in a gloomy frame of mind, wondering if after all I had come to the wrong place. Even the faint whisper of crickets outside in the garden and the huge trembling stars that told me I was back in the tropics did little to cheer me.

The next morning, however, things looked brighter. After breakfast I was out in the garden watching a flock of gold, blue and silver butterflies feeding on the scarlet blooms of a bush when Luna arrived. I had heard him singing, in a pleasant tenor voice, as he came down the avenue of bamboo, and as he reached the gate he paused in his song, clapped his hands in the customary manner of anyone in South America when arriving at your house, opened the gate and joined me by the bush and the butterflies. He was a tiny man, about five feet in height, and as slender as a fourteen-year-old boy. He had a handsome, faintly skull-like face, with huge dark eyes, and black hair that was cropped close to his head. He held out a hand that looked as fragile as the butterflies we were surrounded by.

'Señor Durrell?' he inquired.

'Yes,' I replied, shaking his hand gently, for fear it should break off at the wrist.

'I am Luna,' he said, as if this should be sufficient explanation.

'Señor Lett sent you?' I asked.

'*Si, si,*' he answered, giving me a smile of great charm and sweetness.

We both stood and watched the butterflies drifting round the red blooms, while I racked my brains for the right Spanish phrases.

'*¡Que lindo,*' said Luna, pointing at the butterflies, '*que bichos más lindos!*'

'*Si,*' I said. There was another pause, and we smiled at each other amiably.

'You speak English?' I inquired hopefully.

'No, very small,' said Luna, spreading his hands and smiling gently, as if deploring this terrible gap in his education.

It was obvious that his knowledge of my language was about as extensive as mine was of his. This later proved to be true. Both of us could understand a quite complicated conversation in each other's language, but both were incapable of doing more in speech than string a few ungrammatical nouns and verbs together.

'You . . . I . . . go Helmuth,' suggested Luna suddenly, waving a delicate hand.

I agreed, wondering what a Helmuth was; it was a new word to me, and, as far as I was concerned, could have been anything from a new type of jet-engine to a particularly low night-club. However, I was willing to try anything once, especially if it turned out to be a night-club. We walked down the musically squeaking, creaking, groaning and rustling avenue of bamboo, and then came to a large area of lawn, dotted with gigantic palm trees, their trunks covered with parasitic plants and orchids. We walked through these towards a long, low red brick building, while the humming-birds flipped and whirred around us, gleaming and changing with the delicate sheen one sees on a soap bubble. Luna led me through gauze-covered doors into a large cool dining-room, and there, sitting in solitary state at the end of a huge table, devouring breakfast, was a man of about thirty with barley-sugar coloured hair, vivid blue eyes, and a leathery, red, humorous face. He looked up as we entered and gave us a wide, impish grin.

'Helmuth,' said Luna, pointing to this individual, as if he had performed a particularly difficult conjuring trick. Helmuth rose from the table and extended a large, freckled hand.

'Hullo,' he said, crushing my hand in his, 'I'm Helmuth. Sit down and have some breakfast, eh?'

I explained that I had already had some breakfast, and so Helmuth returned to his victuals, talking to me between mouthfuls, while Luna, seated the other side of the table, drooped languidly in a chair and hummed softly to himself.

'Charles tells me you want animals, eh?' said Helmuth. 'Well, we don't know much about animals here. There *are* animals, of course, up in the hills, but I don't know what you'll get in the villages. Not much, I should think. However, when I've finished eating we go see, eh?'

When Helmuth had assured himself, somewhat reluctantly, that there was nothing edible left on the table, he hustled Luna and myself out to his station-wagon, piled us in and drove down to the village, over the dusty, rutted roads that would, at the first touch of rain, turn into glutinous mud.

The village was a fairly typical one, consisting of small shacks with walls built out of the jagged off-cuts from the saw mill, and whitewashed. Each stood in its own little patch of ground, surrounded by a bamboo fence, and these gardens were sometimes filled with a strange variety of old tins, kettles and broken barrels each brimming over with flowers. Wide ditches full of muddy water separated these 'gardens' from the road, and were spanned at each front gate by a small, rickety bridge of roughly-nailed branches. It was at one of these little shanties that Helmuth stopped. He peered hopefully into the riot of pomegranate trees, covered with red flowers, that filled the tiny garden.

'Here, the other day, I think I see a parrot,' he explained.

We left the station-wagon and crossed the rickety little bridge that led to the bamboo gate. Here we clapped our hands and waited patiently. Presently, from inside the little shack, erupted a brood of chocolate-coloured children, all dressed in clean but tattered clothing, who lined up like a defending army and regarded us out of black eyes, each, without exception, sucking its thumb vigorously. They were followed by their mother, a short, rather handsome Indian woman with a shy smile.

'Enter, señores, enter,' she called, beckoning us into the garden.

We went in, and, while Luna crouched down and conducted a muttered conversation with the row of fascinated children, Helmuth, exuding goodwill and personality, beamed at the woman.

'This señor,' he said, gripping my shoulder tightly, as if fearful that I might run away, 'this señor wants *bichos*, live *bichos*, eh? Now, the other day when I passed your house, I saw that you possessed a parrot, a very common and rather ugly parrot of a kind that I have no doubt the señor will despise. Nevertheless, I am bound to show it to him, worthless though it is.'

The woman bristled.

'It is a beautiful parrot,' she said shrilly and indignantly, 'a very beautiful parrot, and one, moreover, of a kind that is extremely rare. It comes from high up in the mountains.'

'Nonsense,' said Helmuth firmly, 'I have seen many like it in the market in Jujuy, and they were so common they were practically having to give them away. This one is undoubtedly one of those.'

'The señor is mistaken,' said the woman, 'this is a most unusual bird of great beauty and tameness.'

'I do not think it is beautiful,' said Helmuth, and added loftily, 'and as for its tameness, it is a matter of indifference to the señor whether it be tame or as wild as a puma.'

I felt it was about time I entered the fray.

'Er . . . Helmuth,' I said tentatively.

'Yes?' he said, turning to me and regarding me with his blue eyes flashing with the light of battle.

'I don't want to interfere, but wouldn't it be a good idea if I saw the bird first, before we start bargaining? I mean, it might be something very common, or something quite rare.'

'Yes,' said Helmuth, struck by the novelty of this idea, 'yes, let us see the bird.'

He turned and glared at the woman.

'Where is this wretched bird of yours?' he inquired.

The woman pointed silently over my left shoulder, and turning round I found that the parrot had been perching among the green leaves of the pomegranate tree some three feet away, an interested spectator of our bargaining. As soon as I saw it I knew that I must have it, for it was a rarity, a red-fronted Tucuman Amazon, a bird which was, to say the least of it, unusual in European collections. He was small for an Amazon parrot, and his plumage was a rich grass-green with more than a tinge of yellow in it here and there; he had bare white rings round his dark eyes, and the whole of his forehead was a rich scarlet. Where the feathering ended on each foot he appeared to be wearing orange garters. I gazed at him longingly. Then, trying to wipe the acquisitive look off my face I turned to Helmuth and shrugged with elaborate unconcern, which I am sure did not deceive the parrot's owner for a moment.

'It's a rarity,' I said, trying to infuse dislike and loathing for the parrot into my voice, 'I must have it.'

'You see?' said Helmuth, returning to the attack, 'the señor says it is a very common bird, and he already has six of them down in Buenos Aires.'

The woman regarded us both with deep suspicion. I tried to look like a man who possessed six Tucuman Amazons, and who really did not care to acquire any more. The woman wavered, and then played her trump card.

'But this one *talks*,' she said triumphantly.

'The señor does not care if they talk or not,' Helmuth countered quickly. We had by now all moved towards the bird, and were gathered in a circle round the branch on which it sat, while it gazed down at us expressionlessly.

'*Blanco . . . Blanco*,' cooed the woman, '*¿como te vas, Blanco?*'

'We will give you thirty pesos for it,' said Helmuth.

'Two hundred,' said the woman, 'for a parrot that talks, two hundred is cheap.'

'Nonsense,' said Helmuth, 'anyway, how do we know it talks? It hasn't said anything.'

'*Blanco, Blanco*,' cooed the woman in a frenzy, 'speak to Mama . . . speak Blanco.'

Blanco eyed us all in a considering way.

'Fifty pesos, and that's a lot of money for a bird that won't talk,' said Helmuth.

'*Madre de Dios*, but he talks all day,' said the woman, almost in tears, 'wonderful things he says . . . he is the best parrot I have ever heard.'

'Fifty pesos, take it or leave it,' said Helmuth flatly.

'Blanco, Blanco, speak,' wailed the woman, 'say something for the señores . . . please.'

The parrot shuffled his green feathering with a silken sound, put his head on one side and spoke.

'*Hijo de puta*,' he said, clearly and slowly.

The woman stood as though transfixed, her mouth open, unable to believe in the perfidy of her pet. Helmuth uttered a great sigh as of someone who knows the battle is won. Slowly, and with a look of utter malignancy, he turned to the unfortunate woman.

'So!' he hissed, like the villain in a melodrama. 'So! This is your idea of a talking parrot, eh?'

'But señor . . . ,' began the woman faintly.

'Enough!' said Helmuth, cutting her short. 'We have heard enough. A stranger enters your gates, in order to help you by paying you money (which you need) for a worthless bird. And what do you do? You try and cheat him by telling him your bird talks, and thus get him to pay more.'

'But it *does* talk,' protested the woman faintly.

'Yes, *but what does it say*?' hissed Helmuth. He paused, drew himself up to his full height, took a deep breath and roared:

'It tells this good-natured, kindly señor that he is the son of a whore.'

The woman looked down at the ground and twiddled her bare toes in the dust. She was beaten and she knew it.

'Now that the señor knows what disgusting things you have taught this bird I should not think he will want it,' continued Helmuth. 'I should think that now he will not even want to offer you fifty pesos for a bird that has insulted not only him, but his mother.'

The woman gave me a quick glance, and returned to the contemplation of her toes. Helmuth turned to me.

'We have got her,' he said in a pleading tone of voice, 'all you have to do is to try and look insulted.'

'But I am insulted,' I said, trying to look offended and suppress the desire to giggle. 'Never, in fact, in a long career of being insulted, have I been so insulted.'

'You're doing fine,' said Helmuth, holding out both hands as if begging me to relent. 'Now give in a bit.'

I tried to look stern but forgiving, like one of the less humorous saints one sees in ikons.

'All right,' I said reluctantly, 'but only this once. Fifty, you said?'

'Yes,' said Helmuth, and as I pulled out my wallet he turned again to the woman. 'The señor, because he is the very soul of kindness, has forgiven you the insult. He will pay you the fifty pesos that you demanded, in your greed.'

The woman beamed. I paid over the grubby notes, and then approached the parrot. He gazed at me musingly. I held out my finger, and he gravely climbed on to it, and then made his way up my arm to my shoulder. Here he paused, gave me a knowing look, and said quite clearly and loudly:

'*¿Como te vas, como te vas, que tal?*' and then giggled wickedly.

'Come on,' said Helmuth, revitalized by his session of bargaining. 'Let's go and see what else we can find.'

We bowed to the woman, who bowed to us. Then, as we closed the bamboo gate behind us and were getting into the car, Blanco turned on my shoulder and fired his parting shot.

'*Estupido*,' he called to his late owner, '*muy estupida*.'

'That parrot,' said Helmuth, hastily starting the car, 'is a devil.'

I was inclined to agree with him.

Our tour of the village was not entirely unproductive. By careful questioning and cross-questioning nearly everyone we met, we managed to run to earth five yellow-fronted Amazon parrots, an armadillo and two grey-necked guans. These latter are one of the game-birds, known locally as *charatas*, which is an onomatopoeic name resembling their cry. They look, at first glance, rather like a slim and somewhat drab hen pheasant of some species. Their basic colouring is a curious brown (the pale colour a stale bar of chocolate goes) fading to grey on the neck. But, see them in the sun and you discover that what you thought was a matt brown is really slightly iridescent with a golden sheen. Under the chin they have two drooping red wattles, and the feathers on their heads, when they get excited, stand up in a kind of crest that looks like a lengthy crew-cut. They were both young birds, having been taken from the nest when a few days old and hand-reared, so they were ridiculously tame. The Amazon parrots were also tame, but none of them had the knowingness of the vocabulary of Blanco. All they could do was to mutter '*Lorito*' at intervals, and whistle shrilly. Nevertheless, I felt for one morning's work we were not doing too badly, and so I carried my purchases back in triumph to the house, where Joan Lett had kindly allowed me to use their empty garage as a sort of storehouse for my creatures.

As I had no cages ready for the reception of my brood, I had to let them all loose in the garage and hope for the best. To my surprise this arrangement worked very well. The parrots all found themselves convenient perches, just out of pecking range of each other, and, though it had obviously been agreed that Blanco was the boss, there was no unmannerly squabbling. The guans also found themselves perches, but these they only used to sleep on, preferring to spend their days stalking about the floor of the garage, occasionally throwing back their heads and letting forth their ear-splitting cry. The armadillo, immediately on being released, fled behind a large box, and spent all day there meditating, only tip-toeing out at night to eat his food, casting many surreptitious and fearful glances at the sleeping birds.

By the following day the news had spread through the village that there had arrived a mad *gringo* who was willing to pay good money for live animals, and the first trickle of specimens started. The first arrival was an Indian carrying, on the end of a length of string, a coral snake striped in yellow, black and scarlet, like a particularly revolting old school tie. Unfortunately, in his enthusiasm, the Indian had tied the string too tightly about the reptile's neck, and so it was very dead.

I had better luck with the next offering. An Indian arrived clasping a large straw hat tenderly to his bosom. After a polite exchange of greetings I asked to see what he had so carefully secured in his hat. He held it out, beaming hopefully at me, and then looking into the depths of the hat I saw reclining at the bottom, with a dewy-eyed expression on its face, the most delightful kitten. It was a baby Geoffroy's cat, a small species of wild cat which is getting increasingly rare in South America. Its basic colouring was a pale fawny yellow, and it was dappled all over with neat, dark brown spots. It regarded me with large bluey-green eyes from the interior of the hat, as if pleading to be picked up. I should have known better. In my experience it is always the most innocent-looking creatures that can cause you the worst damage. However, misled by its seraphic expression, I reached out my hand and tried to grasp it by the scruff of the neck. The next moment I had a bad bite through the ball of my thumb and twelve deep red grooves across the back of my hand. As I withdrew my hand, cursing, the kitten resumed its innocent pose, apparently waiting to see what other little game I had in store for it. While I sucked my hand like a half-starved vampire, I bargained with the Indian and eventually purchased my antagonist. Then I tipped it, hissing and snarling like a miniature jaguar, out of the hat and into a box full of straw. There I left it for an hour or so to settle down. I felt that its capture and subsequent transportation in a straw hat might be mainly responsible for its fear and consequent bad temper, for the creature was only about two weeks old, as far as I could judge.

When I thought it had settled down and would be willing to accept my overtures of friendship, I removed the lid of the box and peered in hopefully. I missed losing my left eye by approximately three millimetres. I wiped the blood from my cheek thoughtfully; obviously my latest specimen was not going to be easy. Wrapping my hand in a piece of sacking I placed a saucer of raw egg and minced meat in one corner of the box, and a bowl of milk in the other, and then left the kitten to its own devices. The next morning neither of the two offerings of food had been touched. With a premonition that this was going to hurt me more than the kitten, I filled one of my feeding-bottles with warm milk, wrapped my hand in sacking and approached the box.

Now I have had, at one time and another, a fair amount of experience in trying to get frightened, irritated or just plain stupid animals to feed from a bottle, and I thought that I knew most of the tricks. The Geoffroy's kitten proceeded to show me that, as far as it was concerned, I was a mere tyro at the game. It was so lithe, quick and strong for its size that after half an hour of struggling I felt as though I had been trying to pick up a drop of quicksilver with a couple of crowbars. I was covered in milk and blood and thoroughly exhausted, whereas the kitten regarded me with blazing eyes and seemed quite ready to continue the fight for the next three days if necessary. The thing that really irritated me was that the kitten had – as I knew to my cost – very well-developed teeth, and there seemed no reason why it should not eat and drink of its own accord, but, in this stubborn mood, I knew that it was capable of quite literally starving itself to death. A bottle seemed the only way of getting any nourishment down it. I put it back in its box, washed my wounds, and was just applying plaster to the deeper of them when Luna arrived, singing cheerfully.

'Good morning, Gerry,' he said, and then stopped short and examined my bloodstained condition. His eyes widened, for I was still bleeding profusely from a number of minor scratches.

'What's this?' he asked.

'A cat . . . *gato*,' I said irritably.

'Puma . . . jaguar?' he asked hopefully.

'No,' I said reluctantly, '*chico gato montes*.'

'*Chico gato montes*,' he repeated incredulously, 'do this?'

'Yes. The bloody little fool won't eat. I tried it on the bottle, but it's just like a damned tiger. What it really needs is an example . . .' my voice died away as an idea struck me.

'Come on, Luna, we'll go and see Edna.'

'Why Edna?' inquired Luna breathlessly as he followed me down the road to Helmuth's flat.

'She can help,' I said.

'But, Gerry, Helmuth won't like it if Edna is bitten by a *gato montes*,' Luna pointed out in Spanish.

'She won't get bitten,' I explained. 'I just want her to give me a kitten.'

Luna gazed at me with dark, puzzled eyes, but the conundrum was too much for him, and so he merely shrugged and followed me round to Helmuth's front door. I clapped my hands and went into Helmuth's and Edna's comfortable sitting-room, where Edna was ensconced over a huge pile of socks, darning placidly and listening to the gramophone.

'Hullo,' she said, giving us her wide, attractive smile, 'the gin is over there, help yourself.'

Edna had a beautiful and placid nature: nothing seemed to worry her unduly. I am sure that if you walked into her sitting-room with fourteen Martians in tow she would merely smile and point out the location of the gin.

'Thank you dear,' I said, 'but I didn't come for gin, strange though it may sound.'

'It does sound strange,' agreed Edna, grinning at me mischievously. 'Well, if you don't want gin, what do you want?'

'A kitten.'

'A kitten?'

'Yes . . . you know, a small cat.'

'Today Gerry is *loco*,' said Luna with conviction, pouring out two liberal measures of gin and handing one to me.

'I have just bought a baby *gato montes*,' I explained to Edna. 'It's extremely wild. It won't eat by itself, and this is what it did to me when I tried to feed it on the bottle.' I displayed my wounds. Edna's eyes widened.

'But how big is this animal?' she asked.

'About the size of a two-week-old domestic cat.'

Edna looked stern. She folded up the sock she was darning.

'Have you put disinfectant on those cuts?' she inquired, obviously preparing herself for a medical orgy.

'Never mind the cuts . . . I washed them . . . But what I want from you is a kitten, an ordinary kitten. Didn't you say the other day that you were infested with kittens over here?'

'Yes,' said Edna, 'we have plenty of kittens.'

'Good. Well, can I have one?'

Edna considered.

'If I give you a kitten will you let me disinfect your cuts?' she asked cunningly. I sighed.

'All right, blackmailer,' I said.

So Edna disappeared into the kitchen quarters, from whence came a lot of shrill exclamations and much giggling. Then Edna returned with a bowl of hot water and proceeded to minister unto my cuts and bites, while a procession of semi-hysterical Indian maids filed into the room, carrying in their arms groups of kittens of all shapes and colours, from ones still blind to ones that were half-grown and looked almost as wild as my Geoffroy's cat. Eventually I chose a fat, placid female tabby which was approximately the same size and age as my wild cat, and carried it back in triumph to the garage. Here I spent an hour constructing a rough cage, while the tabby kitten purred vigorously and rubbed itself round my legs, occasionally tripping me up. When the cage was ready I put the tabby kitten in first, and left it for an hour or so to settle down.

Most wild animals have a very strong sense of territory. In the wild state, they have their own particular bit of forest or grassland which they consider their own preserve, and will defend it against any other member of their own species (or other animals sometimes) that tries to enter it. When you put wild animals into cages the cages become, as far as they are concerned, their territory. So, if you introduce another animal into the same cage, the first inmate will in all probability defend it vigorously, and you may easily have a fight to the death on your hands. So you generally have to employ low cunning. Suppose, for example, you have a large vigorous creature who is obviously quite capable of looking after itself, and it has been in a cage for a period of a few weeks. Then you get a second animal of the same species, and you want to confine them together, for the sake of convenience. Introduce the new specimen into the old one's cage, and the old one may well kill it. So the best thing to do is to build an entirely new cage, and into this you introduce the weaker of the two animals. When it has settled down, you then put the stronger one in with it. The stronger one will, of course, still remain the dominant animal, and may even bully the weaker one, but as far as he is concerned he had been introduced into someone else's territory, and this takes the edge off his potential viciousness. It's a sort of Lifemanship that any collector has to practise at one time or another.

In this case I was sure that the baby Geoffroy's was quite capable of killing the domestic kitten, if I introduced the kitten to *it*, instead of the other way round. So, once the tabby had settled down, I seized the Geoffroy's and pushed it, snarling and raving, into the cage, and stood back to see what would happen. The tabby was delighted. It came forward to the angry Geoffroy's and started to rub itself against its neck, purring loudly. The Geoffroy's, taken aback by its greeting as I had hoped, merely spat rather rudely, and retreated into a corner. The tabby, having made the first overtures of friendship, sat down, purring loudly, and proceeded to wash itself with a self-satisfied air. I covered the front of the cage with a piece of sacking and left them to settle down, for I was sure now that the Geoffroy's would do the tabby no real harm.

That evening, when I lifted the sacking, I found them lying side by side, and the Geoffroy's, instead of spitting at me as it had done up until now, contented itself with merely lifting its lip in a warning manner. I carefully inserted a large bowl of milk into the cage, and a plate containing the finely chopped meat and raw egg, which I wanted the Geoffroy's to eat. This was the crucial test, for I was hoping that the tabby would fall upon this delicious fare and, by example, encourage the Geoffroy's to eat. Sure enough, the tabby, purring like an ancient outboard engine, flung itself at the bowl of milk, took a long drink and then settled down to the meat and egg. I had retreated to a place where I could see without being seen, and I watched the Geoffroy's carefully. To begin with it took no interest at all, lying there with half-closed eyes. But eventually the noise the tabby was making over the egg and meat – it was a rather messy feeder – attracted its attention. It rose cautiously and approached the plate, while I held my breath. Delicately it sniffed round the edge of the plate, while the tabby lifted a face that was dripping with raw egg and gave a mew of encouragement, slightly muffled by the portion of meat it had in its mouth. The Geoffroy's stood pondering for a moment, and then, to my delight, sank down by the plate and started to eat. In spite of the fact that it must have been extremely hungry it

ate daintily, lapping a little raw egg, and then picking up a morsel of meat which it chewed thoroughly before swallowing. I watched them until, between them, they had cleaned both plates, then I replenished them with more milk, egg and meat, and went to bed well satisfied. The next morning both plates were spotless, and the Geoffroy's and the tabby were locked in each other's arms, fast asleep, their stomachs bulging like two little hairy balloons. They did not wake up until midday, and then they both looked distinctly debauched. But when they saw me approaching with the plates of food they both displayed considerable interest, and I knew that my battle with the Geoffroy's was won.

6.

A City of Bichos

The excitement from the novelty of objects, and the chance of
success, stimulate him to increased activity.

CHARLES DARWIN: THE VOYAGE OF H.M.S. BEAGLE

Ever since my arrival in Calilegua, Luna had been pestering me
to accompany him to a town called Oran, which lay some fifty
miles away, and where, he assured me, I would get plenty of
bichos. I was a bit chary about this idea, for I knew how easy it
is to rush frantically from one place to another on a collecting
trip, and, though each place in itself might be a good centre,
you achieve very little by virtue of your grasshopper-like activ-
ities. I decided to discuss it with Charles, and so, that evening,
as we sat gently imbibing gin and watching a moon with a blue
halo silvering the palm fronds, I put my problem to him.

'Why is Luna so keen on Oran?' I asked.

'Well,' said Charles drily, 'it's his home town, for one thing, but this might prove an advantage, for it means that he knows everyone. I think you could do worse than go and investigate, Gerry. It's got a much bigger population than Calilegua, and in view of what you've found here, I should think you'd get twice as much stuff there.'

'Can Luna get the time off?' I asked.

Charles smiled his gentle smile.

'I don't think that we would notice his absence for three days,' he said, 'and that should give you time to denude Oran of whatever fauna is lurking there.'

'Could we leave on Monday?' I inquired hopefully.

'Yes,' said Charles, 'that would be all right.'

'Wonderful,' I said, finishing my drink, 'and now I must go across and see Edna.'

'Why Edna?'

'Well, someone's got to feed my animals while I'm away, and I'm hoping Edna has a kind heart.'

I found Helmuth, Edna and Luna arguing over the relative merits of two folk-songs which they kept playing over and over again on the gramophone. Edna pointed silently to the drinks and I helped myself, and then went and sat on the floor at her feet.

'Edna,' I said, during a lull in the argument, 'I love you.'

She raised one eyebrow sardonically and regarded me.

'If Helmuth wasn't bigger than me I would suggest that we elope,' I went on, 'since the first day I saw you I have been mad about you, your eyes, your hair, the way you pour gin . . .'

'What do you want?' she inquired.

I sighed.

'You have no soul,' I complained. 'I was just getting into my stride. Well, if you must know, Charles says that Luna and I can go to Oran for three days. Will you look after my animals for me?'

'But, of course,' she said, surprised that there should have been any doubt in my mind.

'But, of course,' echoed Helmuth. 'Gerry, you are very stupid. I tell you we will help all we can. You have only to ask. We will try and do anything for you.'

He splashed more gin into my glass.

'Except,' he added reluctantly, 'let you elope with my wife.'

So, early on Monday morning, Luna and I set out in a small station-wagon driven by a gay, semi-inebriated individual, sporting a moustache so large it looked like a Nature Reserve. We took with us only the bare essentials of travel: Luna's guitar, three bottles of wine, my wallet well stuffed with pesos, recording machine and cameras. We also had a clean shirt each, which our driver had placed reverently and tenderly in a pool of oil. All the previous night it had rained with a loudness and thoroughness that only the tropics can achieve; this now had thinned out to a fine grey drizzle, but the earth road had turned into something resembling the consistency of a badly-made blancmange. Luna, undeterred by the weather, the surface of the road and the doubtful driving capabilities of our driver, the fate of our clean shirts and the fact that the roof of the station-wagon leaked daintily but persistently, sang happily to himself as we slithered and swooped along the road to Oran.

We had been travelling some three-quarters of an hour when our driver, concentrating more on harmonizing with Luna in a mournful song than on the car, rounded a corner on two wheels, and as we slithered miraculously on to the straight again I saw something ahead that made my heart sink. Before us lay a torrent of red, froth-flecked water some four hundred yards across. At the edge of this, like a line of depressed elephants, stood three lorries, while in mid-stream, twisted to one side by the force of the water, another lorry was being laboriously dragged across to the opposite bank by a thing like a gigantic tractor, fitted with a winch and steel cable. Our driver joined the line of waiting lorries, switched off his engine and beamed at us.

'*Mucho agua,*' he pointed out to me, in case my eyesight should be defective and I had missed noticing the miniature Bay of Biscay we had to cross. I knew that the previous day this broad torrent had probably been a mere trickle of water, shallow and glinting over its bed of pebbles, but one night's rain had swollen it suddenly and out of all proportion. I knew, from experience, how a tiny stream can grow into a fierce full-sized river in next to no time, for once in West Africa I had had my camp almost washed away by a stream that started by being a mere three feet wide and four inches deep, and had, in the course of an hour or so, turned into something resembling the upper reaches of the Amazon. No one who has not seen this sudden transformation can believe it, but it can be one of the most irritating (and sometimes dangerous) aspects of travel in the tropics.

At last, after an hour of waiting, the last of the lorries had been hauled over and it was our turn. The hawser was attached to our bumper and gingerly we were drawn into the flood. Slowly the water rose higher and higher, and became stronger, until it was rustling and lapping along one side of the station-wagon like a miniature tidal wave. The water spurted in through the cracks of the door and trickled across the floor under our feet. Gradually the water rose until it covered our shoes. We were now approximately half-way across, and the force of the water was kindly but firmly pushing us downstream so that, although to begin with we had been opposite the tractor and the winch, we were now some fifty yards downstream from them. The hawser was taut, and I felt as though we were some gigantic and misshapen fish that the two laconic-looking Indians on the tractor were playing. The water had now reached the level of the seats; here it paused for a moment and then overflowed generously under our behinds. At this crucial moment, sitting in half an inch of icy water, we heard the winch stop.

'Arrrr!' roared our driver, sticking his head out of the window, his moustache quivering impressively, '*¿que pasa?*'

One of the Indians leapt off the tractor, and loped slowly off down the road; the other pushed his big straw hat on to the back of his head and slowly approached the bank of the river.

'*Nafta no hay,*' he explained, scratching his stomach with every evidence of satisfaction.

'Fine bloody time for them to run out of petrol,' I said irritably to Luna.

'Yes,' said Luna despondently, 'but the other Indian has gone for some. He will not be long.'

Half an hour passed. Then an hour. By now our nether regions were so frozen that we were all shifting uneasily in our seats to try and get some feeling back, making noises like a troupe of hippopotami enjoying a wallow in a particularly succulent swamp. At last, to our relief, the Indian appeared loping down the road carrying a can of petrol. He and the other Indian then had a long argument as to the best method of putting the life-giving fluid into the tractor, while our driver roared insults at them from between chattering teeth. But at last they had finished this highly complicated operation, the tractor sprang into life, the hawser tightened and we were drawn slowly but inexorably towards the bank, while the water-level in the wagon fell.

When we eventually reached dry land we all got out, removed our trousers and wrung them out, while our driver soundly berated the Indians for their attempted homicide, while they both grinned amiably at us. Then the driver, in his shirt-tails, opened the bonnet of the car and peered into the engine, his moustache twitching, muttering to himself. He had carefully wrapped in cotton waste certain vital parts of the internal organs of our vehicle before we entered the flood, and these he now unwrapped, and then proceeded to dry other parts of the engine. Eventually, he climbed in, pressing the starter, and with a wide grin of pride, heard the engine roar into life. We piled in and jolted off down the road, the Indians waving their straw hats in gay farewell.

We had travelled some five miles and were just beginning to dry out when we met our next water hazard. The road here ran along the lower slopes of the mountains, and the terrain was intersected at intervals by deep, narrow gorges through which the water from the mountains drained. Where the road crossed one of these narrow but powerful rivers one would have thought that the simplest engineering method would have been to throw a small bridge across from bank to bank. Apparently the vast numbers of these rivers made this too costly, and so another method was employed. A slightly concave apron of cement was laid across the river bed, which at least gave your wheels some purchase. In the dry season, of course, this looked merely like a continuation of the road, but when the waters from the mountains stormed down they roared over the apron, sometimes four feet deep, and then dropped into a graceful ten foot waterfall the other side to join the river lower down. A few days covered with water and the cement developed a surface like glass, owing to the algae that adhered to it, and so it was considerably more hazardous than the original river-bed would have been.

Here there was no winch to help us, and the driver nosed the station-wagon carefully into the red water, scowling fearfully behind his bristling moustache. We had got half-way across the invisible cement apron, when the engine stalled. We sat and looked at each other mutely, until suddenly the force of the water piling up against the side of the vehicle shifted it an inch or so in the direction of the waterfall on our right, and then we were all suddenly galvanized into activity. We none of us wanted to be sitting in the station-wagon if the torrent suddenly got a good grip on her and swept her over the edge and downstream among the tangle of rocks we could see. We left the vehicle as one man.

'Push . . . we must all push,' said Luna, raising his voice above the noise of the falls. He was clinging to the side of the station-wagon with both hands, for the force of the water was considerable. He was so slight in build that I expected at any moment to see him plucked away by the current and swept over the waterfall like a feather.

'Go round the other side of the car,' I shouted, 'the water won't sweep you away there.'

Luna realized the force of this argument, and made his way round the wagon in a starfish-like manner, until it stood between him and the waterfall. Then we laid our shoulders to the wagon and started to push. It was quite one of the most unrewarding tasks I have ever undertaken, for not only were we trying to push the wagon up the opposite slope of the cement apron, but we were also pushing against the current which all the time was trying to twist the wagon round at an angle. After about ten minutes of struggling we had managed to shift our vehicle approximately three feet nearer the opposite bank, and the current had moved it three feet nearer the edge of the waterfall. I began to get really worried, for at this rate I could see the wagon plunging gracefully over the waterfall in a matter of another half-hour, for the three of us alone had not the strength to push her up the slope *and* against the current. We had a rope in the back of the wagon and, if it was long enough, the only thing I could suggest was that we tethered the wagon to a tree on the opposite bank, and just sat there until the waters subsided. I was just about to try and put this plan into Spanish, when round the corner of the road on the opposite bank appeared a Fairy Godmother, heavily disguised as a wheezing, snorting lorry, which, in spite of its age and rust, looked powerful and phlegmatic. We greeted it with shouts of joy. The driver of the lorry took in our predicament in a glance, and slowing down, drove the vast bulk of his vehicle slowly into the red torrent until he was within a few feet of us. Hastily we got out our rope and shackled the two vehicles together; then the lorry went into reverse and gently drew our vehicle out of the flood and on to dry land. We thanked the lorry driver, gave him a cigarette, and watched enviously as he drove his mighty steed through the torrent as if it had not been there. Then we turned our attention to the laborious and messy process of drying out our engine.

Eventually we reached Oran at two o'clock in the afternoon, having had to navigate three more water hazards, none of which, fortunately, was as bad as the last two. Nevertheless, we arrived at Luna's house all looking as though we had spent our entire day in the river, which was not so far from the truth. Luna's charming family greeted us with delight, whipped our clothes away to be dried, cooked us an enormous meal, and sat us down to eat it in an indoor courtyard, overflowing with flowers, where the frail sunlight was just starting to make its heat felt. While we ate and drank good, warming red wine, Luna sent an apparently endless stream of his smaller relatives on mysterious missions to different parts of the town, and they kept reappearing to whisper reports to him, whereupon he would nod his head portentously and smile, or else scowl ferociously, according to the news that was being vouchsafed to him. Everyone had an air of suppressed excitement, and stiffened expectantly if Luna so much as coughed or looked in their direction. I began to feel as though I was having lunch with the Duke of Wellington on the eve of Waterloo. At last he leant forward, poured us both out a last glass of wine, and then grinned at me, his big black eyes sparkling with suppressed excitement.

'Gerry,' he said in Spanish, 'I have found you some *bichos.*'

'Already?' I asked. 'But how?'

He waved a hand at his small army of relatives, standing in a grinning line.

'I have sent my family to make inquiries, and they have discovered a number of people who have *bichos.* Now it only remains for us to go and buy them if they are the *bichos* you want.'

'Wonderful,' I said enthusiastically, finishing my wine at a gulp, 'let's go, shall we?'

So, in ten minutes' time, Luna and I set off to quarter Oran like huntsmen, preceded by our pack of Luna's young and excited relatives. The town was not really so large, but rather straggling, built on the typical Argentine chessboard pattern. Everywhere we went, as Charles had predicted, Luna was greeted with cries of joy, and we had to refuse many invitations of the more bibulous variety. But Luna, with a reluctant gleam in his eye, sternly turned his back on such frivolity, and we continued on our way. Eventually, one of the younger members of our retinue ran on ahead and beat a loud tattoo on a most impressive-looking door of a large house. By the time we had reached it the door had been opened by an ancient woman dressed in black, which made her look like a somewhat dilapidated cockroach. Luna paused in front of her and gave her a grave good evening, to which she bowed slightly.

'I know that you have in your house a parrot,' said Luna with the air of a policeman daring a criminal to deny the existence of a corpse which he knows to be concealed beneath the sofa.

'That is so,' said the woman, mildly surprised.

'This English señor is collecting for his *jardin zoologico* in England,' Luna went on, 'and it is possible that he may wish to purchase this bird of yours.'

The woman surveyed me from dark, dry eyes, without curiosity.

'You are welcome to him,' she said at last, 'for he is a dirty bird and he does not talk. My son brought him to me, but if I can sell him I will be only too glad. Come in, señores, and see him.'

She shuffled ahead of us and led us into the inevitable courtyard of potted plants, forming the well of the house. When I saw the bird it was all I could do to stifle a yelp of delight, for the creature was a yellow-naped macaw, a rare member of the parrot family. It was sitting on the remains of a wooden perch which it had obviously, over the past week, demolished slowly and systematically until scarcely anything remained. It glanced up at us as we gathered round it, a fine sliver of wood in its beak, uttered a short gurking noise, and returned to its work of demolition. Luna gave me a quick glance from his brilliant eyes, and I nodded my head vigorously. He took a deep breath, surveyed the macaw with loathing, and then turned to the woman.

'One of the commoner ones, I see,' he said carelessly, 'but even so the señor is interested in buying it. You realize, of course, that for such a common, destructive bird, and one, moreover, that does not talk, we cannot afford to be generous. The señor would not dream of considering paying anything more than, say, twenty-five pesos for such a creature.'

Then he folded his arms and looked at the woman, waiting for her outburst of indignation at the mere mention of such a low price.

'All right,' said the woman, 'you can have him.'

While Luna regarded her open-mouthed she picked up the macaw, plonked him unceremoniously on my shoulder, and held out her wrinkled palm for the notes which I was hastily counting out from my wallet before she changed her mind. We were back in the street again, with the macaw making surprised and pleased gurking noises in my ear, before Luna recovered the power of speech. Then he shook his head despondently.

'What's the matter, Luna?' I asked. 'It's a wonderful bird, and to get it so cheap is incredible.'

'For your sake,' said Luna gloomily, 'I am glad. But it makes me fear for the future of Argentina when I meet someone who will not bargain, but accepts the first price offered. Where would we all be if everyone did that?'

'Life would probably be a lot cheaper,' I pointed out, but he refused to be comforted, and continued to grumble over the woman's behaviour for the rest of our tour of the town, though a brisk half-hour exchange with a man who drove a hard bargain over another parrot shortly restored his faith in humanity.

We continued on our way through the town until it grew dark, by which time all of us were carrying what amounted to a small zoo. There were five parrots (including, to my delight, another yellow-naped macaw), two pigmy Brazilian rabbits, with ginger paws and white spectacles of fur round their eyes, and an orange-rumped agouti, a large rodent with dark eyes, slender legs and the disposition of a racehorse suffering from an acute nervous breakdown. We carried this assortment of wild-life back to Luna's house and let them all loose in the patio, while Luna organized his band of relatives once more and sent them scurrying in all directions to fetch empty boxes, wire-netting, saws, hammers, nails and other accoutrements of the carpenter's trade. Then, for the next two hours we were fully occupied building suitable habitations for my acquisitions. At length, when the last of the creatures had been placed in its cage, Luna and I sat at the table nearby and ate and drank heartily, while from the pile of wooden boxes came the faint scufflings and squawks which are such music to the animal collector's ears. Presently, a large tumbler of good wine by my side, I sat down in front of the cages to examine my charges by lamplight, while Luna called for his guitar and sang the soft, mournful folk-songs of Argentina, occasionally, where the music required it, using the deep wooden belly of his guitar as a drum.

The parrots we had acquired were all blue-fronted Amazons, all rather scruffy because of bad feeding, but all reasonably tame and able to mutter the inevitable '*Lorito*' which is the Argentine equivalent of 'Polly'. As they were all much the same size and age we had caged them together, and now in the lamplight they sat in a row, like a highly coloured jury, regarding me with the ancient, reptilian and falsely-wise expressions that parrots are such masters at adopting. I was pleased with them in spite of their tattered appearance, for I knew that a few weeks of good feeding would make a world of difference, and that, at their next moult, their feathers would glow with lemon-yellow, blue and a multitude of greens that would make a collection of emeralds look dowdy in comparison. Gently I lowered a piece of sacking over the front of their cage and heard them all fluff and rearrange their feathers (a sound like someone riffling through a pack of cards) preparatory to sleep. Next I turned my attention to the yellow-naped macaws, and gloated over them for some time. We had, experimentally, caged them together, and the way they had immediately taken to each other and started to bill and coo inclined me to think that they were a true pair. They sat on the perch now and regarded me solemnly, occasionally turning their heads on one side as if to see whether I looked any more attractive that way. Basically their colouring was a deep, rush green, only relieved on the neck where they had a broad half-moon-shaped patch of feathers which were bright canary-yellow. For macaws – which are as a rule the largest of the parrots – they were diminutive, being slightly smaller and more slender than the common Amazon parrots. They gurked gently to me and to each other, their pale eyelids drooping sleepily over their bright eyes, so I covered them up with sacking and left them.

Next to the macaws the Brazilian rabbits were the creatures I was most delighted to have obtained, for they were animals I had long wanted to meet. The two we had got were only babies, and I lifted them out of their cage and they sat, one in each hand, comfortably filling my palms with the soft, fat warmth of their bodies, their noses wiffling with all the strange scents of food and flowers with which the patio was filled. At first glance you would have taken them for the young of the common European rabbit, but closer inspection soon showed the differences. To begin with their ears were very short for their size, and very neat and slender. The basic colouring on the back was a dark rich brown, flecked and patterned with rusty-coloured patches and blobs. Their feet and part of the leg were a bright, rich ginger, and, as I said before, they had a fine circle of white hair round each eye. Their nose and lips, I now noticed, were faintly outlined in white as well. When they were fully adult, I knew, they would still be among the dwarfs of their breed, being only half the size of the European wild rabbit. As far as I knew, no zoo in the world possessed these interesting little creatures, and I was delighted to

have got them, though I had faint qualms about being successful in taking them back to Europe, for the rabbit and hare family do not, on the whole, take kindly to captivity, and are reputedly difficult. However, these were very young, and I had hopes that they would settle down satisfactorily.

When I lifted the sacking off the front of the agouti's cage she leapt straight up into the air, and landed with a crash in her straw bed, quivering in every limb, with the expression of an elderly virgin who, after years of looking under the bed, has at last found a man there. However, with the aid of a piece of apple I managed to soothe her into a fairly reasonable state, and she actually allowed me to stroke her. Agoutis are, of course, rodents, members of that enormous and interesting family that includes creatures like the harvest mouse, which would hardly fill the bowl of a teaspoon, to capybaras that are the size of a large dog and in between these two extremes a great variety of squirrels, dormice, rats, porcupines and other unlikely beasts. Agoutis are not, let us admit at once, the most prepossessing of their family. To be perfectly frank, they look like a cross between one of the smaller forerunners of the horse, and a rather lugubrious rabbit. Their basic colouring is a rich, shining mahogany, fading to reddish-ginger on their rumps. Their legs are chocolate brown, very long and slender and racehorse-like, ending in a bunch of frail, artistic toes which give them the ancient-horse look. Their hind legs are powerful in order to support a backside that is out of all proportion to the forequarters, so that the creature looks, if I might put it like this, as though it had a hump-behind. The head is rabbit-like, but again slightly elongated so that there is still a faint suggestion of horse about it. They have large, fine eyes, neat rounded ears and a mass of black whiskers which are in a constant state of agitation about everything. Combine all this with the beast's temperament, its constantly neurotic state, its wild leaps into the air at the slightest sound followed by a period of acute ague, and you begin to wonder how the species survives at all. I should imagine that a jaguar would only have to growl once and every agouti within a hundred yards' radius would die of heart-failure immediately. Musing on this I lowered the sacking over the front of my agouti's cage, and she immediately leapt once more into the air and came down shaking in every limb. However, within a few minutes she had recovered from this terrible experience sufficiently to make an attack on the apple I left in the cage for her. Luna had now, by the application of song and wine, worked himself into a pleasant state where he sat at the table, humming softly like a drowsy bee. We had a final glass of wine as a nightcap, and then yawning prodigiously, stumbled off to bed.

I was awoken at what seemed to me to be a most uncivilized hour of the morning by a burst of song from Luna's bed, in the opposite corner of the room. Song and music ran through Luna's being as naturally as the blood flowed through his veins. When he was not talking he was singing or humming, and he is the only man I have ever met who can stay up until three in the morning and rise at five, bursting into song before he is even out of bed. But he sang so pleasantly and with such obvious pleasure that you forgave him, even at that hour of the morning, and, after knowing him some time, you took no more notice of it than you would have done of a dawn chorus of birds.

'The moon is a like a little white drum in the sky,' he sang from under a pile of bedclothes, 'leading me to my love with the dark hair and the magic eyes, behind the mountains of Tucuman.'

'If you sing to all your female acquaintances at this hour of the morning,' I said drowsily, 'I should think you lead a pretty lonely life in bed. These things get around you know.'

He chuckled and stretched luxuriously.

'Today is going to be a fine day, Gerry,' he said. I wondered how he knew, for the shutters on the two windows were tightly closed. The night air, in which the Argentine will sit as late as he pleases without any harm to his being, becomes, as soon as he retires to bed, a deadly gas waiting to strangle him. So all shutters must be tightly closed to guard against such a danger-ous experience. However, when we had dressed and gone out into the patio to breakfast, I found he was right, for it was flooded with sunshine.

We were finishing our last cup of coffee when our troupe of spies appeared to report. Apparently they had been out and about at the crack of dawn, and they made their reports to Luna as he sat there, sipping his coffee, and occasionally giving a lordly nod. Then one of the younger of the spies was dispatched with money to purchase provisions for my specimens, and, on his return, the spies stood wide-eyed and watched me while I chopped up food and vegetables, filled bowls with milk or water, and generally min-istered to my animals. When the last one had been fed, we filed out into the sunlit street and started once more on our search of the town. This time Luna used our retinue slightly differently. While we made our way to a house which we knew contained some wild pet, our young helpers fanned out and explored every alley and street in the immediate vicinity, clapping their hands outside people's doors, and questioning complete strangers as to what pets they kept. Everyone was most good-humoured about this intrusion of their privacy, and, even if they had no creatures themselves, would sometimes direct us to another house in which lurked some member of the local fauna. By this means, during the morning, we ran to earth three more pigmy rabbits, another parrot, two seriemas, a strange, leggy type of bird, and two coatimundis, the odd little raccoon-like predator of South America. We took them back to Luna's house, caged them, ate a hearty lunch and then, exhilarated by our morning's success, set out to explore the outer limits of Oran, with the aid of an ancient car, lent to us by one of Luna's friends.

Luna had learnt, by some M.I.5 methods of his own, that in one of the more far-flung portions of the town was a man who possessed a wild cat of some sort, but no one was quite sure who it was or the exact location of his house. Eventually, however, we narrowed our search down to one rambling street, and by the simple process of knocking or clapping outside every house we eventually found the man we were looking for. He was a large, dark, sweating and unclean-looking man of about forty, with an unhealthy paunch and beady black eyes that were alternately cringing or cunning. Yes, he admitted, he had got a wild cat, an ocelot; and then, with all the fiery eloquence of a pre-election politician, he proceeded to tell us about the animal's beauty, grace, tameness, value, coloration, size, appetite, until I began to feel that he was trying to sell me an entire zoo instead of one animal. Breaking in on his asthmatic eulogy on the feline tribe in general and his specimen in particular, we asked to see it. He led us round into one of the filthiest backyards I had been in to date, for in Oran and in Calilegua, however poor and tiny the house, the backyard was always neat and full of flowers. This looked like a council rubbish dump, with old broken barrels, rusty tin cans, piles of old wire-netting, bicycle wheels and other flotsam and jetsam. Our host lumbered over to a rough wooden cage in one corner which would have been small for the average rabbit. He opened the door, caught hold of a chain inside and hauled out on to the ground one of the most pathetic sights I have seen. It was a half-grown ocelot, and how it managed to fit in such a small cage was a mystery. But it was its condition that was so appalling. Its coat was so matted with its own filth that you could only just discern the natural pattern of the skin. It had a large, running sore on one flank, and it was so thin that you could, under its matted coat, see its ribs and backbone clearly. Indeed, it was so weak that it wavered from side to side, like a drunk, when it was dropped on to the ground, and eventually gave up the attempt to stay upright, and sank dejectedly down on to its dirty belly.

'You see how tame it is?' inquired the man, giving us a display of tattered yellow teeth in an ingratiating grin. 'She is very tame with everybody. Never has she been known to bite.' He was patting the cat as he spoke, with one great sweaty palm. I could see that it was not tameness that stopped the animal from turning on him, but sheer inertia due to lack of food. She had almost reached the point of no return, where she felt so weak that she just did not care.

'Luna,' I said, making a valiant attempt to keep my temper, 'I will pay fifty pesos for this cat. No more. Even that is too much, for she will probably die. I won't bargain, so you can tell this bloated illegitimate son of an inadequate whore that he can take it or leave it.'

Luna translated my message, tactfully leaving out my character rendering. The man clasped his hands in horror. Surely we were joking? He giggled feebly. For such a magnificent animal three hundred pesos would be a beggarly sum to pay. Surely the señor could see what a wonderful creature . . . and so on. But the señor had seen enough. I spat loudly and accurately into the remains of a barrel, lovingly entwined with a bicycle-wheel, gave the man the dirtiest look I could achieve, turned on my heel and walked back to the road. I got into our ancient car and slammed the door with such violence that, for a moment, I thought the whole thing was going to fall to pieces in the road. I could hear Luna and the man arguing, and presently, when I detected a weakening note in the repulsive man's voice, I leant out of the window and roared at Luna to come on and not waste time. Within thirty seconds he appeared.

'Give me the money, Gerry,' he said. I gave him the fifty pesos, and presently he appeared with the box and put it in the back seat. We drove off in silence. Presently, when I had finished mentally working out what I would like to have done to the cat's late owner, which would not only have been painful but have made his marital state, if any, difficult in the extreme, I sighed and lit a cigarette.

'We must get home quickly, Luna. That animal's got to have a decent cage and some food or she's going to die,' I said. 'Also I shall want some sawdust.'

'*Si, si,*' said Luna, his dark eyes worried. 'I have never seen anyone keep an animal like that. She is half-dead.'

'I think I can save her,' I said. 'At least, I think we've got a fifty-fifty chance.'

We drove in silence along the rutted road for some way before Luna spoke.

'Gerry, you do not mind stopping once more, only for a minute?' he inquired anxiously. 'It is on our way. I hear of someone else that has a cat they might sell.'

'Yes, all right, if it's on the way. But I hope to God it's in better condition than the one we've got.'

Presently Luna ran the car off the road on to a sizeable stretch of greensward. On one corner of this stood a dilapidated-looking marquee, and near it a small, battered-looking merry-go-round and a couple of small booths made of striped canvas now so faded as to be almost white. Three fat, glossy horses, one a bright piebald, grazed near by, and around the marquee and the booths trotted a number of well-fed-looking dogs, who had the air of professionals.

'What is this? It looks like a circus,' I said to Luna.

'It is a circus,' said Luna, grinning, 'only a very small one.'

I was amazed that any circus, even a small one, could make a living in a place as remote and small as Oran, but this one appeared to be doing all right for, although the props were some-what decrepit, the animals looked in good condition. As we left the car a large ginger-haired man appeared, ducking out from under the flap of the marquee. He was a muscular individual with shrewd green eyes and powerful, well-kept hands, who looked as though he would be capable of doing a trapeze act or a lion act with equal skill. We shook hands, and Luna explained our business.

'Ah, you want my puma,' he grinned. 'But I warn you I want a lot of money for her . . . she's a beauty. But she eats too much, and I can't afford to keep her. Come and see her, she's over here. A real devil, I can tell you. We can't do a thing with her.'

He led us to a large cage in one corner of which crouched a beautiful young puma, about the size of a large dog. She was fat and glossy, and still had her baby paws which, as in all young cats, look about three times too big for the body. Her coat was a rich amber colour, and her piercing, moody eyes a lovely leaf green. As we approached the cage she lifted one lip and showed her well-developed baby teeth in a scornful snarl. She was simply heavenly, and a joy to look at after the half-starved creature we had just bought, but I knew, fingering my wallet, that I should have to pay a lot for her.

The bargaining lasted for half an hour and was conducted over a glass of very good wine which the circus proprietor insisted we drank with him. At length I agreed to a price which, though high, seemed to me to be fair. Then I asked the man if he would keep her until the following day for me, if I paid for her evening meal, for I

knew that she would be in good hands, and I had no cage ready for her reception. This our amiable ginger friend agreed to and the bargain was sealed with another glass of wine, and then Luna and I drove back home to try and resurrect the unfortunate ocelot.

When I had built a cage for her, and one of Luna's lesser relatives had appeared with a large sackful of sweet-smelling sawdust, I got the poor creature out of her evil-smelling box and dressed the wound on her thigh. She just lay on the ground apathetically, though the washing of the wound must have hurt considerably. Then I gave her a large shot of penicillin, which again she took no notice of. The third operation was to try and dry her coat out a bit, for she was drenched with her own urine, and already the skin of her belly and paws were fiery red, burnt by the acid. All I could do was literally to cover her in sawdust, rubbing it well into the fur to absorb the moisture, and then gently dusting it out again. Then I unpicked the more vicious tangles in her fur, and by the time I had finished she had begun to look faintly like an ocelot. But she still lay on the floor, uncaring. I cut the filthy collar away from her neck, and put her in her new cage on a bed of sawdust and straw. Then I placed in front of her a bowl containing one raw egg and a small quantity of finely-minced fresh steak. At first she displayed no interest in this, and my heart sank, for I thought she might well have reached the stage of starvation where no amount of tempting offerings would induce her to eat. In sheer desperation I seized her head and ducked her face into the raw egg, so that she would be forced to lick it off her whiskers. Even this indignity she suffered without complaint, but she sat back and licked the dripping egg off her lips, slowly, carefully, like someone sampling a new, foreign and probably dangerous dish. Then she eyed the dish with a disbelieving look in her eye. I honestly think that the animal, through ill-treatment and starvation, had got into a trance-like state, where she disbelieved the evidence of her own senses. Then, while I held my breath, she leant forward and lapped experimentally at the raw egg. Within thirty seconds the plate was clean, and Luna and I were dancing a complicated tango of delight round the patio, to the joy of his younger relatives.

'Give her some more, Gerry,' panted Luna, grinning from ear to ear.

'No, I daren't,' I said. 'When a creature's that bad you can kill it from overfeeding. She can have a bowl of milk later on, and then tomorrow she can have four small meals during the day. But I think she'll be all right now.'

'That man was a devil,' said Luna, shaking his head.

I drew a deep breath and, in Spanish, gave him my views on the cat's late owner.

'I never knew you knew so many bad things in Spanish, Gerry,' said Luna admiringly. 'There was one word you used I have never heard before.'

'I've had some good teachers,' I explained.

'Well, I hope you say nothing like that tonight,' said Luna, his eyes gleaming.

'Why? What's happening tonight?'

'Because we are leaving tomorrow for Calilegua, my friends have made an *asado* in your honour, Gerry. They will play and sing only very old Argentine folk-songs, so that you may record them on your machine. You like this idea?' he asked anxiously.

'There is nothing I like better than an *asado*,' I said, 'and an *asado* with folk-songs is my idea of Heaven.'

So, at about ten o'clock that evening, a friend of Luna's picked us up in his car and drove us out to the estate, some distance outside Oran, where the *asado* had been organized. The *asado* ground was a grove near the *estancia*, an area of bare earth that told of many past dances, surrounded by whispering eucalyptus trees and massive oleander bushes. The long wooden benches and trestle-tables were lit with the soft yellow glow of half a dozen oil-lamps, and outside this buttercup circle of light the moonlight was silver brilliant. There were about fifty people there, many of whom I had never met, and few of them over the age of twenty. They greeted us uproariously, almost dragged us to the trestle-tables which were groaning under the weight of food, and placed great hunks of steak, crisp and sizzling from the open fires, in front of us. The wine bottles passed with monotonous regularity, and within half-an-hour Luna and I were thoroughly in the party spirit, full of good food, warmed with red wine. Then these gay, pleasant young people gathered round while I got the recorder ready, watching with absorbed attention the mysteries of thread-ing tape and getting levels. When, at last, I told them I was ready, guitars, drums and flutes appeared as if by magic, and the entire crowd burst into song. They sang and sang, and each time they came to the end of a song, someone would think of a new one, and they would start again. Sometimes a shy, grinning youth would be pushed to the front of the circle as the only person there capable of rendering a certain number, and after much encouragement and shouts of acclamation he would sing. Then it would be a girl's turn to sing the solo refrain in a sweet-sour voice, while the lamps glinted on her dark hair, and the guitars shuddered and trembled under the swiftly-moving brown fingers of their owners. They danced in a row on a flagstoned path, their spurs ringing sparks from the stone, so that I could record the heel-taps which are such an intricate part of the rhythm of some of their songs; they danced the delightful handkerchief dance with its pleasant lilting tune, and they danced tangos that made you wonder if the stiff, sexless dance called by that name in Europe was a member of the same family. Then, shouting with laughter because my tapes had run out and I was in despair, they rushed me to the table, plied me with more food and wine, and sitting round me sang more sweetly than ever. These, I say again, were mostly teenagers, revelling in the old and beautiful songs of their country, and the old and beautiful dances, their faces flushed with delight at my delight, honouring a stranger they had never seen before and would probably never see again.

By now they had reached the peak. Slowly they started to relax, the songs getting softer and softer, more and more plaintive, until we all reached the moment when we knew the party was over, and that to continue it longer would be a mistake. They had sung themselves from the heavens back to earth, like a flock of descending larks. Flushed, bright-eyed, happy, our young hosts insisted that we travelled back to Oran with them in the big open back of the lorry in which they had come. We piled in, our tightly-packed bodies creating a warmth for which we were grateful, for the night air was now chilly. Then as the lorry roared off down the road to Oran, bottles of red wine were passed carefully from hand to hand, and the guitarists started strumming. Everybody, revived by the cool night air, took up the refrain, and we roared along through the velvet night like a heavenly choir. I looked up and saw the giant bamboos that curved over the road, now illuminated by the lorry's headlights. They looked like the talons of some immense green dragon, curved over the road, ready to pounce if we stopped singing for an instant. Then a bottle of wine was thrust into my hand, and as I tipped my head back to drink I saw that the dragon had passed, and the moon stared down at me, white as a mushroom-top against the dark sky.

7.

Vampires and Wine

The vampire bat is often the cause of much trouble, by biting the horses on their withers.

CHARLES DARWIN: THE VOYAGE OF H.M.S. BEAGLE

On my return from Oran the garage almost overflowed with animals. One could scarcely make oneself heard above the shrill, incomprehensible conversations of the parrots (interspersed occasionally with a shrill scream, as if some local female was being raped), the harsh rattling cries of the guans, the incredibly loud trumpeting song of the seriemas, the chittering of the coati-mundis, and an occasional dull rumble, as of distant thunder, from the puma, whom I had christened Luna in the human Luna's honour. As a background to this there was a steady scrunching noise that came from the agouti cage, for it was always engaged in trying to do alterations to its living quarters with its chisel-like teeth.

As soon as I had got back I had begun constructing cages for all our various creatures, leaving the caging of Luna until last, for she had travelled in a large packing-case that gave her more than enough room to move about in. However, when all the other animals were housed, I set about building a cage worthy of the puma, which, I hoped, would show off her beauty and grace. I had just finished it when Luna's godfather arrived, singing lustily as usual, and offered to help me in the tricky job of getting Luna to pass from her present quarters into the new cage. We carefully closed the garage doors so that, if anything untoward happened, the cat would not go rampaging off across the countryside and be lost. It also had the advantage, as the human Luna pointed out, that we would be locked in with her, a prospect he viewed with alarm and despondency. I soothed his fears by telling him that the puma would be far more frightened than we were, and at that moment she uttered a rumbling growl of such malignance and fearlessness that Luna paled visibly. My attempt to persuade him that this growl was really an indication of how afraid the animal was of us was greeted with a look of complete disbelief.

The plan of campaign was that the crate in which the puma now reposed would be dragged opposite the door of the new cage, a few slats removed from the side, and the cat would then walk from the crate into the cage without fuss. Unfortunately owing to the somewhat eccentric construction of the cage I had built, we could not wedge the crate close up to the door: there was a gap of some eight inches between crate and cage. Undeterred, I placed planks so that they formed a sort of short tunnel between the two boxes, and then proceeded to remove the end of the crate so that the puma could get out. During this process a golden paw, that appeared to be the size of a ham, suddenly appeared in the gap and a nice, deep slash appeared across the back of my hand.

'Ah!' said Luna gloomily, 'you see, Gerry?'

'It's only because she's scared of the hammering,' I said with feigned cheerfulness, sucking my hand. 'Now, I think I've removed enough boards for her to get through. All we have to do is wait.'

We waited. After ten minutes I peered through a knot hole and saw the wretched puma lying quietly in her crate, drowsing peacefully, and showing not the slightest interest in passing down our rickety tunnel and into her new and more spacious quarters. There was obviously only one thing to do, and that was to frighten her into passing from crate to cage. I lifted the hammer and brought it down on the back of the crate with a crash. Perhaps I should have warned Luna. Two things happened at once. The puma, startled out of her half-sleep, leapt up and rushed to the gap in the crate, and the force of my blow with the hammer knocked down the piece of board which was forming Luna's side of the tunnel. In consequence he looked down just in time to see an extremely irritable-looking puma sniffing meditatively at his legs. He uttered a tenor screech which I have rarely heard equalled, and leapt vertically into the air. It was the screech that saved the situation. It so unnerved the puma that she fled into the new cage as fast as she could, and I dropped the sliding door, locking her safely inside. Luna leant against the garage door wiping his face with a handkerchief.

'There you are,' I said cheerfully, 'I told you it would be easy.'

Luna gave me a withering look.

'You have collected animals in South America and Africa?' he inquired at length. 'That is correct?'

'Yes.'

'You have been doing this work for fourteen years?'

'Yes.'

'You are now thirty-three?'

'Yes.'

Luna shook his head, like a person faced with one of the great enigmas of life.

'How you have lived so long only the good God knows,' he said.

'I lead a charmed life,' I said. 'Anyway, why did you come to see me this morning, apart from wanting to wrestle with your namesake?'

'Outside,' said Luna, still mopping his face, 'is an Indian with a *bicho*. I found him with it in the village.'

'What kind of *bicho*?' I asked as we left the garage and went out into the garden.

'I think it is a pig,' said Luna, 'but it's in a box and I can't see it very clearly.'

The Indian was squatting on the lawn, and in front of him was a box from which issued a series of falsetto squeaks and muffled grunts. Only a member of the pig family could produce such extraordinary sounds. The Indian grinned, removed his big straw hat, ducked his head, and then, removing the lid of the box, drew forth the most adorable little creature. It was a very young collared peccary, the common species of wild pig that inhabits the tropical portions of South America.

'This is Juanita,' said the Indian, smiling as he placed the diminutive creature on the lawn, where it uttered a shrill squeak of delight and started to snuffle about hopefully.

Now, I have always had a soft spot for the pig family, and baby pigs I cannot resist, so within five minutes Juanita was mine at a price that was double what she was worth, speaking financially, but only a hundredth part of what she was worth in charm and personality. She was about eighteen inches long and about twelve inches high, clad in long, rather coarse greyish fur, and a neat white band that ran from the angle of her jaw up round her neck, so that she looked as though she was wearing an Eton collar. She had a slim body, with a delicately tapered snout ending in a delicious retroussé nose (somewhat like a plunger), and slender, fragile legs tipped with neatly polished hooves the circumference of a sixpence. She had a dainty, lady-like walk, moving her legs very rapidly, her hooves making a gentle pattering like rain.

She was ridiculously tame, and had the most endearing habit of greeting you – even after only five minutes' absence – as if you had been away for years, and that, for her, these years had been grey and empty. She would utter strangled squeaks of delight, and rush towards you, to rub her nose and behind against your legs in an orgy of delighted reunion, giving seductive grunts and sighs. Her idea of Heaven was to be picked up and held on her back in your arms, as you would nurse a baby, and then have her tummy scratched. She would lie there, her eyes closed, gnashing her baby teeth together, like miniature castanets, in an ecstasy of delight. I still had all the very tame and less destructive creatures running loose in the garage, and as Juanita behaved in such a lady-like fashion I allowed her the run of the place as well, only shutting her in a cage at night. At feeding time it was a weird sight to see Juanita, her nose buried in a large dish of food, sur-rounded by an assortment of creatures – seriemas, parrots, pigmy rabbits, guans – all trying to feed out of the same dish. She always behaved impeccably, allowing the others plenty of room to feed, and never showing any animosity, even when a wily seriema pinched titbits from under her pink nose. The only time I ever saw her lose her temper was when one of the more weak-minded of the parrots, who had worked himself into a highly excitable state at the sight of the food plate, flew down squawking joy-ously, and landed on Juanita's snout. She shook him off with a grunt of indignation and chased him, squawking and fluttering, into a corner, where she stood over him for a moment, champing her teeth in warning, before returning to her interrupted meal.

When I had got all my new specimens nicely settled, I paid a visit to Edna to thank her for the care and attention she had lavished on my animals in my absence. I found her and Helmuth busy with a huge pile of tiny scarlet peppers, with which they were concocting a sauce of Helmuth's invention, an ambrosial substance which,

when added to soup, removed the roof of your mouth with the first swallow, but added a flavour that was out of this world. An old boot, I am sure, boiled and then covered with Helmuth's sauce, would have been greeted with shouts of joy by any gourmet.

'Ah, Gerry,' said Helmuth, rushing to the drink cupboard, 'I have got good news for you.'

'You mean you've bought a new bottle of gin?' I inquired hopefully.

'Well, that of course,' he said, grinning. 'We knew you were coming back. But apart from that do you know that next week-end is a holiday?'

'Yes, what about it?'

'It means,' said Helmuth, sloshing gin into glasses with gay abandon, 'that I can take you up the mountains of Calilegua for three days. You like that, eh?'

I turned to Edna.

'Edna,' I began, 'I love you . . .'

'All right,' she said resignedly, 'but you must make sure the puma can't get out, that is all I insist upon.'

So, the following Saturday morning, I was awoken, just as dawn was lightening the sky, by Luna, leaning through my window and singing a somewhat bawdy love-song. I crawled out of bed, humped my equipment on to my back, and we made our way through the cool, aquarium light of dawn to Helmuth's flat. Outside it was a group of rather battered-looking horses, each clad in the extraordinary saddle that they use in the north of Argentina. The saddle itself had a deep, curved seat with a very high pommel in front, so that it was almost like an armchair to sit in. Attached to the front of the saddle were two huge pieces of leather, shaped somewhat like angel's wings, which acted as wonderful protection for your legs and knees when you rode through thorn scrub. In the dim dawn light the horses, clad in these weird saddles, looked like some group of mythical beasts, Pegasus for example, grazing forlornly on the dewy grass. Nearby lounged a group of four guides and hunters who were to accompany us, delightfully wild and unshaven-looking, wearing dirty *bombachas*, great wrinkled boots and huge, tattered straw hats. They were watching Helmuth, his corn-red hair gleaming with dew, as he rushed from horse to horse, stuffing various items into the sacks which were slung across the saddles. These sacks, Helmuth informed me, contained our rations for the three days we should be away. Peering into two of the sacks I discovered that our victuals consisted mainly of garlic and bottles of red wine, although one sack was stuffed with huge slabs of unhealthy-looking meat, the blood from which was dripping through the sacking, and whose curious shape gave one the rather unpleasant impression that we were transporting a dismembered body. When everything was to Helmuth's satisfaction, Edna came out, shivering in her dressing gown, to see us off, and we mounted our bony steeds and set off at a brisk trot towards the mountain range which was our goal, dim, misty and flecked here and there with gold and green in the morning light.

At first we rode along the rough tracks that ran through the sugar-cane fields, where the canes whispered and clacked in the slight breeze. Our hunters and guides had cantered on ahead, and Luna and I and Helmuth rode in a row, keeping our horses at a gentle walk. Helmuth was telling me the story of his life, how, at the age of seventeen (as an Austrian), he had been press-ganged into the German Army, and had fought through the entire war, first in North Africa, then Italy and finally in Germany, without losing anything except the top joint of one finger, which was removed by a land-mine that blew up under him and should have killed him. Luna merely slouched in his great saddle, like a fallen puppet, singing softly to himself. When Helmuth and I had settled world affairs generally, and come to the earth-shaking conclusion that war was futile, we fell silent and listened to Luna's soft voice, the chorus of canes, and the steady clop of our horses' hooves, like gentle untroubled heart-beats in the fine dust.

Presently the path left the cane fields and started to climb up the lower slopes of the mountains, passing into real forest. The massive trees stood, decorated with trailing epiphytes and orchids, each one bound to its fellow by tangled and twisted lianas, like a

chain of slaves. The path had now taken on the appearance of an old watercourse (and in the rainy season I think this is what it must have been) strewn with uneven boulders of various sizes, many of them loose. The horses, though used to the country and sure-footed, frequently stumbled and nearly pitched you over their heads, so you had to concentrate on holding them up unless you suddenly wanted to find yourself with a split skull. The path had now narrowed, and twisted and turned through the thick undergrowth so tortuously that, although the three of us were riding almost nose to tail, we frequently lost sight of each other, and if it had not been for Luna's voice raised in song behind me, and the occasional oaths from Helmuth when his horse stumbled, I could have been riding alone. We had been riding this way for an hour or so, occasionally shouting comments or questions to each other, when I heard a roar of rage from Helmuth, who was a fair distance ahead. Rounding the corner I saw what was causing his rage.

The path at this point had widened, and along one side of it ran a rock-lined ravine, some six feet deep. Into this one of our pack horses had managed to fall, by some extraordinary means known only to itself, for the path at this point was more than wide enough to avoid such a catastrophe. The horse was standing, looking rather smug I thought, in the bottom of the ravine, while our wild-looking hunters had dismounted and were trying to make it climb up on to the path again. The whole of one side of the horse was covered with a scarlet substance that dripped macabrely, and the animal was standing in what appeared to be an ever-widening pool of blood. My first thought was to wonder, incredulously, how the creature had managed to hurt itself so badly with such a simple fall, and then I realized that the pack that the horse was carrying contained, among other things, part of our wine supply. The gooey mess and Helmuth's rage were explained. We eventually got the horse back up on to the path, and Helmuth peered into the wine-stained sack, uttering moans of anguish.

'Bloody horse,' he said, 'why couldn't it fall on the *other* side, where the meat is?'

'Anything left?' I asked.

'No,' said Helmuth, giving me an anguished look, 'every bottle broken. Do you know what that means, eh?'

'No,' I said truthfully.

'It means we have only twenty-five bottles of wine to last us,' said Helmuth. Subdued by this tragedy we proceeded on our way slowly. Even Luna seemed affected by our loss, and sang only the more mournful songs in his extensive repertoire.

We rode on and on and on, the path getting steeper and steeper. At noon we dismounted by a small, tumbling stream, our shirts black with sweat, bathed ourselves and had a light meal of raw garlic, bread and wine. This, to the fastidious, may sound revolting, but when you are hungry there is no finer combination of tastes. We rested for an hour, to let our sweat-striped horses dry off, and then mounted again and rode on throughout the afternoon. At last, when the evening shadows were lengthening and we could see glimmers of a golden sunset through tiny gaps in the trees above, the path suddenly flattened out, and we rode into a flat, fairly clear area of forest. Here we found that our hunters had already dismounted and unsaddled the horses, while one of them had gathered dry brushwood and lighted a fire. We dismounted stiffly, unsaddled our horses and then, using our saddles and the woolly sheepskin saddle-cloth, called a *recado*, as backrests we relaxed round the fire for ten minutes, while the hunters dragged out some of the unsavoury-looking meat from the sacks and set it to roast on wooden spits.

Presently, feeling a bit less stiff, and as there was still enough light left, I decided to have a walk round the forest in the immediate area of our camp. Very soon the gruff voices of the hunters were lost among the leaves as I ducked and twisted my way through the tangled, sunset-lit undergrowth. Overhead an occasional humming-bird flipped and purred in front of a flower for a last-night drink, and small groups of toucans flapped from tree to tree, yapping like puppies, or contemplating me with heads on one side, wheezing like rusty hinges. But it was not the birds that interested me so much as the extraordinary variety of fungi that I saw around me. I have never, in any part of the world, seen such a variety of mushrooms and toadstools littering the forest floor, the fallen tree-trunks, and the trees themselves. They were in all colours, from wine-red to black, from yellow to grey, and in a fantastic variety of shapes. I walked slowly for about fifteen minutes in the forest, and in that time I must have covered an area of about an acre. Yet in that short time, and in such a limited space, I filled my hat with twenty-five different species of fungi. Some were scarlet, shaped like goblets of Venetian glass on delicate stems; others were filigreed with holes, so that they were like little carved ivory tables in yellow and white; others were like great, smooth blobs of tar or lava, black and hard, spreading over the rotting logs, and others appeared to have been carved out of polished chocolate, branched and twisted like clumps of miniature stag's antlers. Others stood in rows, like red or yellow or brown buttons on the shirt-fronts of the fallen trees, and others, like old yellow sponges, hung from the branches, dripping evil yellow liquid. It was a Macbeth witches' landscape, and at any moment you expected to see some crouched and wrinkled old hag with a basket gathering this rich haul of what looked like potentially poisonous fungi.

Soon, it became too dark to see properly between the trees, and I made my way back to camp, spread out my fungi in rows, and examined them by the firelight. The unsavoury-looking meat had by now turned into the most delicious steaks, brown and bubbling, and we each with our own knife kept leaning forward cutting any delicate slivers away from the steaks, dunking them in Helmuth's sauce (a bottle of which he had thoughtfully brought with him) and popping the fragrant result into our mouths. Except for an occasional belch the silence was complete. The wine was passed silently, and occasionally someone would lean forward and softly rearrange the logs on the fire, so that the flames flapped upwards more brightly, and the remains of the steaks sizzled briefly, like a nest of sleepy wasps. At last, surfeited with food, we lay back against the comfortable hummocks of our saddles, and Luna, after taking a deep pull at the wine bottle, picked up his guitar and started to strum softly. Presently, very gently, he started to sing, his voice scarcely travelling beyond the circle of firelight, and the hunters joined him in a deep, rich chorus. I put on my poncho (that invaluable garment like a blanket with a hole in the middle), wrapped myself tightly in it – with one hand free to accept the wine bottle as it drifted round the circle – rolled my sheepskin *recado* into a warm, comfortable pillow and lay back, listening to the haunting songs, and watching a white moon edge its way very slowly through the black fretwork of branches above our heads. Then, suddenly, without any preliminary drowsiness, I was deeply asleep.

I awoke, still staring up into the sky, which was now a pale blue, suffused with gold. Turning on my side I saw the hunters already up, the fire lit, and more strips of meat hung to cook. Helmuth was crouching by the fire drinking a huge mug of steaming coffee, and he grinned at me as I yawned.

'Look at Luna,' he said, gesturing with his cup, 'snoring like a pig.'

Luna lay near me, completely invisible under his poncho. I extricated my leg from my own poncho and kicked vigorously at what I thought was probably Luna's rear end. It was, and a yelp greeted my cruelty. This was followed by a giggle and a burst of song as Luna's head appeared through the hole in his poncho, making him look ridiculously like a singing tortoise emerging from its shell. Presently, warmed by coffee and steaks, we saddled up and rode off into the forest, damp and fragrant with dew, and alive with ringing bird-calls.

As we rode my mind was occupied with the subject of vampire bats. I realized that, in the short time at our disposal up the mountains, we had little chance of catching any really spectacular beasts, but I knew that our destination was infested with these bats. At one time an attempt had been made to start a coffee plantation up where we were going, but no horses could be kept because of the vampires, and so the project had been abandoned. Now, I was extremely keen to meet a vampire on its home ground, so to speak, and, if possible, to catch some and take them back to Europe with me, feeding them on chicken's blood, or, if necessary, my own or that of any volunteers I could raise. As far as I knew they had never been taken back to any European zoo, though some had been kept successfully in the United States. I only hoped that, after being so long neglected, all the vampires at the coffee farm had not moved on to more lucrative pastures.

Our destination, when we reached it an hour or so later, proved to be a dilapidated one-roomed hut, with a small covered verandah running along one side. I gave it approximately another six months before it quietly disintegrated and became part of the forest: we had obviously only just arrived in time. All the hunters,

Helmuth and Luna, treated this hut as though it was some luxury hotel, and eagerly dragged their saddles inside and argued amicably over who should sleep in which corner of the worm-eaten floor. I chose to sleep out on the verandah, not only because I felt it would be a trifle more hygienic, but from there I could keep an eye on the tree to which the horses were tethered, for it was on them that I expected the bats to make their first attack.

After a meal we set off on foot into the forest, but, although we saw numerous tracks of tapir and jaguar and lesser beasts, the creatures themselves remained invisible. I did manage, however, by turning over every rotten log we came across, to capture two nice little toads, a tree-frog and a baby coral snake, the latter much to everyone's horror. These I stowed away carefully in the linen bags brought for the purpose when we returned to our hut for the evening meal. When we had finished we sat round the glowing remains of the fire, and Luna, as usual, sang to us. Then the rest of them retired into the hut, carefully closing the window and the door so that not a breath of deadly night air should creep in and kill them (though they had slept out in it quite happily the night before), and I made up my bed on the verandah, propped up so that I could get a good view of the horses, silvered with moonlight, tethered some twenty feet away. I settled myself comfortably, lit a cigarette, and then sat there straining my eyes into the moonlight for the very first sign of a bat anywhere near our horses. I sat like this for two hours before, against my will, dropping off to sleep.

I awoke at dawn, and, furious with myself for having slept, I struggled out of my poncho and went to inspect the horses. I discovered, to my intense irritation, that two of them had been attacked by vampires while I lay snoring twenty feet away. They had both been bitten in exactly the same place, on the neck about a hand's length from the withers. The bites themselves consisted of two even slits, each about half an inch long and quite shallow. But the effect of these small bites was quite gruesome, for the blood (as in all vampire bites) had not clotted after the bat had finished licking up its grisly meal and flown off, for the vampire's saliva contains an anti-coagulant. So, when the bloated bat had left its perch on the horses' necks the wounds had continued to bleed, and now the horses' necks were striped with great bands of clotted blood, out of all proportion to the size of the bites. Again I noticed that the bites, as well as being in identical positions on each horse, were also on the same side of the body, the right side of the animal if you were sitting on it, and there was no sign of a bite or an attempted bite on the left side of either horse. Both animals seemed quite unaffected by the whole thing, and seemed mildly surprised at the interest I was taking in them.

After breakfast, determined that the vampire bats must be lurking somewhere nearby, I organized the rest of the party in a search. We spread out and hunted through the forest in a circle round the hut, going about a quarter of a mile deep into the forest, looking for hollow trees or small caves where the vampires might be lurking. We continued in this fruitless task until lunchtime, and when we reassembled at the hut the only living specimens we could really be said to have acquired were some three hundred and forty black ticks of varying ages and sizes, who, out of all of us, seemed to have preferred the smell of Luna and Helmuth, and so had converged on them. They had to go down to the stream nearby and strip; then, having washed the more tenacious ticks off their bodies, they set about the task of removing the others from the folds and crannies of their clothing, both of them perched naked on the rocks, picking at their clothes like a couple of baboons.

'Curious things, ticks,' I said conversationally, when I went down to the stream to tell them that food was ready, 'parasites of great perception. It's a well-known natural history fact that they always attack the more unpleasant people of the party . . . usually the drunks, or the ones of very low mentality or morals.'

Luna and Helmuth glared at me.

'Would you,' inquired Helmuth interestedly, 'like Luna and me to throw you over that waterfall?'

'You must admit it's a bit peculiar. None of our hunters got ticks, and they are all fairly good parasite-bait, I would have thought. *I* didn't get ticks. You two were the only ones. You know the old English proverb about parasites?'

'What old English proverb?' asked Helmuth suspiciously.

'Birds of a feather flock together,' I said, and hurried back to camp before they could get their shoes on and follow me.

The sun was so blindingly hot in the clearing when we had finished eating that everyone stretched out on the minute veran-dah and had a siesta. While the others were all snoring like a covey of pigs, I found I could not sleep. My head was still full of vampires. I was annoyed that we had not found their hideout, which I felt sure must be somewhere fairly near. Of course, as I realized, there may have been only one or two bats, in which case looking for their hideout in the local forest was three times as dif-ficult as the usual imbecile occupation of looking for needles in haystacks. It was not until the others had woken, with grunts and yawns, that an idea suddenly occurred to me. I jumped to my feet and went inside the hut. Looking up I saw, to my delight, that the single room had a wooden *ceiling*, which meant that there must be some sort of loft between the apex of the roof and the ceiling. I hurried outside and there, sure enough, was a square opening which obviously led into the space between roof and ceiling. I was now convinced that I should find the loft simply stuffed with vampire bats, and so I waited impatiently while the hunters fash-ioned a rough ladder out of saplings and hoisted it up to the hole. Then I sped up it, armed with a bag to put my captures in and a cloth to catch them with without being bitten. I was followed by Helmuth who was going to guard the opening with an old shirt of mine. Eagerly, holding a torch in my mouth, I wriggled into the loft. The first discovery I made was that the wooden ceiling on which I was perched was insecure in the extreme, and so I had to spread myself out like a starfish to distribute my weight, unless I wanted the whole thing to crash into the room below, with me on it. So, progressing on my stomach in the manner of a stalking Red Indian, I set out to explore the loft.

The first sign of life was a long, slender tree-snake, which shot past me towards the hole that Helmuth was guarding. When I informed him of this and asked him to try and catch it he greeted this request in the most unfriendly manner, interspersed with a number of rich Austrian oaths. Luckily for him, the snake found a crack in the ceiling and disappeared through that, and we did not see it again. I crawled on doggedly, disturbing three small scorpions, who immediately rushed into the nearest holes, and eight large and revolting spiders of the more hirsute variety, who merely shifted slightly when the torch-beam hit them, and crouched there meditatively. But there was not the faintest sign of a bat, not even so much as a bat dropping to encourage me. I was just beginning to feel very bitter about bats in general and vampire bats in particular, when my torch-beam picked out something sitting sedately on a cross-beam, glaring at me ferociously, and I immediately forgot all about vampires.

Squatting there in the puddle of torchlight was a pigmy owl, a bird little bigger than a sparrow, with round yellow eyes that glared at me with all the silent indignation of a vicar who, in the middle of the service, has discovered that the organist is drunk. Now, I have a passion for owls of all sorts, but these pigmy owls are probably my favourites. I think it is their diminutive size combined with their utter fearlessness that attracts me; at any rate I determined to add the one perching above me to my collection, or die in the attempt. Keeping the torch beam firmly fixed on his eyes, so that he could not see what I was doing, I gently brought up my other hand and then, with a quick movement, I threw the cloth I carried over him, and grabbed. He uttered a squeak of indignation, and fluttered wildly, sinking his small but sharp talons into my fingers through the cloth. Placing the torch on the

floor I wrapped him up tightly in the cloth and then put the whole bundle inside my shirt and buttoned it up for further safety. Then, having made quite sure once more that there was not a bat in the loft, I started to make my way back to the entrance. This was, to say the least, difficult, for the owl was reposing against my chest, so I had to travel on my back. This gave me a wonderful view of the spiders overhead, all of which now seemed to be the size of soup-plates and each ready to drop on me if I made a false move. Fascinating as I find spiders, I prefer to keep the larger and more hairy varieties at a distance. At last I reached the opening and levered myself out into the sunshine.

To my surprise the hunters were excited and delighted with my capture of the pigmy owl. I was puzzled by this, until they explained that it was a common belief in Argentina that if you possessed one of these little birds you would be lucky in love. This answered a question that had been puzzling me for some time. When I had been in Buenos Aires I had found one of these owls in a cage in the local bird market. The owner had asked a price that was so fantastic that I had treated it with ridicule, until I realized that he meant it. He refused to bargain, and was quite unmoved when I left without buying the bird. Three days later I had returned, thinking that by now the man would be more amenable to bargaining, only to find that he had sold the owl at the price he had asked for. This had seemed to me incredible, and I could not for the life of me think of a satisfactory explanation. But now I realized I had been outbid by some lovesick swain; I could only hope that the owl brought him luck.

That night was to be our last spent in the mountains, and I was grimly determined that I was going to catch a vampire bat if one showed so much as a wing-tip that night. I had even decided that I would use myself as bait. Not only would it bring the bats within catching range, but I was interested to see if the bite was really as painless as it was reputed to be. So, when the others had retired to their airless boudoir, I made up my bed as near to the horses as I felt I could get, without frightening off the bats, wrapped myself up in my poncho but left one of my feet sticking out, for vampires, I had read, were particularly fond of human extremities, especially the big toe. Anyway, it was the only extremity I was prepared to sacrifice for the sake of Science.

I lay there in the moonlight, glaring at the horses, while my foot got colder and colder. I wondered if vampires liked frozen human big toe. Faintly from the dark forest around came the night sounds, a million crickets doing endless carpentry work in the undergrowth, hammering and sawing, forging miniature horseshoes, practising the trombone, tuning harps, and learning how to use tiny pneumatic drills. From the tree-tops frogs cleared their throats huskily, like a male chorus getting ready for a concert. Everything was brilliantly lit by moonlight, including my big toe, but there was not a bat to be seen.

Eventually, my left foot began to feel like something that had gone with Scott to the Pole, and had been left there, so I drew it into the warmth of the poncho and extended my right foot as a sacrifice. The horses, with drooping heads, stood quite still in the moonlight, very occasionally shifting their weight from one pair of legs to another. Presently, in order to get some feeling back into my feet, I went and hobbled round the horses, inspecting them with the aid of a torch. None of them had been attacked. I went back and continued my self-imposed torture. I did a variety of things to keep myself awake: I smoked endless cigarettes under cover of the poncho, I made mental lists of all the South American animals I could think of, working through the alphabet, and, when these failed and I started to feel sleepy, I thought about my overdraft. This last is the most successful sleep eradicator I know. By the time dawn had started to drain the blackness out of the sky, I was wide awake and feeling as though I was solely responsible for the National Debt. As soon as it became light enough to see without a torch I hobbled over to inspect the horses, more as a matter of form than anything. I could hardly believe my eyes, for two of them were painted with gory ribbons of blood down their necks. Now, I had been watching those horses – in brilliant moonlight – throughout the night, and I would have staked my life that not a bat of any description had come within a hundred yards of them. Yet two of them had been feasted upon, as it were, before my very eyes. To say that I was chagrined is putting it mildly. I had feet that felt as if they would fall off at a touch, a splitting headache, and felt generally rather like a dormouse that had been pulled out of its nest in mid-October.

Luna and Helmuth, of course, when I woke them up, were very amused, and thought this was sufficient revenge for my rude remarks the previous day about parasites. It was not until I had finished my breakfast in a moody and semi-somnambulistic state, and was starting on my third mug of coffee, that I remembered something that startled me considerably. In my enthusiasm to catch a vampire bat, and to be bitten by one to see what it felt like, I had completely forgotten the rather unpleasant fact they can be rabies carriers, so being bitten by one might have had some interesting repercussions, to say the least. I remembered that the rabies vaccine (which, with the usual ghoulish medical relish, they inject into your stomach) is extremely painful, and you have to have a vast quantity of the stuff pumped into you before you are out of danger. Whether this is necessary, or simply because the doctors get a rake-off from the vaccine manufacturers, I don't know, but I do know – from people that have had it – that is it not an experience to be welcomed. The chances of getting rabies from a bat in that particular area would be extremely slight, I should have thought, but even so, had I been bitten, I would have had to undergo the injections as a precautionary measure; anyone who has ever read a description of the last stages of a person suffering from rabies would be only too happy to rush to the nearest hospital.

So, without bats or bites, and with my precious pigmy owl slung round my neck in a tiny bamboo cage, we set off down the mountains back to Calilegua. By the time we reached the cane fields it was green twilight, and we were all tired and aching. Even Luna, riding ahead, was singing more and more softly. At length we saw the glow of lights from Helmuth's flat, and when we dismounted, stiff, sweaty and dirty, and made our way inside, there was Edna, fresh and lovely, and by her side a table on which stood three very large ice-cold gin-and-tonics.

8.

A Wagon-Load of Bichos

In conclusion, it appears to me that nothing can be more improving to a young naturalist than a journey in distant countries.

CHARLES DARWIN: THE VOYAGE OF H.M.S. BEAGLE

I have, during the course of my perambulations about the world, met a number of curious and interesting human beings. If I were to make a list of these characters the top places would be occupied by two people I met during my last ten days at Calilegua.

Helmuth came in one morning and informed me that he had to travel to an *estancia* some distance away from Calilegua, for some purpose or other, and that near his destination was another *estancia* run by a man who (he had been informed) kept animals as pets. So while he conducted his business he would drop me off at the other *estancia* to see what bargains I could discover. As we drove along Helmuth told me something about the man I was to meet.

'I have never seen him, but all the local people say that he comes from a very good European family. They say he used to entertain kings and princes when his father was chief minister of one of the Balkan states. I don't know how much is true . . . you know what it's like in this sort of place, Gerry, eh? They say anything about your past life, and if they can't think of anything about your past life, they simply say that you are not married to your wife, or that you are a drunk or a homosexual, or something of the sort.'

'Yes, I know what it's like,' I said, 'I once lived in a cosy English village where you couldn't pass the time of day with any female between the ages of seven and seventy without being accused of rape.'

'Still,' said Helmuth philosophically, 'if he has *bichos* for you, who cares what he is.'

After driving for a couple of hours we turned off the road and made our way along bumpy tracks through sugar-cane fields. Presently we came to a pleasant, small, one-storeyed house, surrounded by a well-tended garden. The lawn was scattered with the evidence of children: a rocking-horse, a battered teddy bear, a rough cement paddling-pool with a small yacht listing heavily to starboard floating in it.

'Here we are,' said Helmuth. 'Out you get. I'll pick you up in a couple of hours' time. O.K.?'

'O.K.,' I agreed, as I got out. 'What's this chap's name, anyway?'

'Caporal,' said Helmuth, and drove off, leaving me standing in a cloud of dust.

Here I might say that the person in question was not called Caporal, but I have not used his real name for a variety of reasons, the chief one being that he did not give me permission to write about him. Anyway, when I had sneezed some of the dust out of my nose, I clapped dutifully outside the gate, and then opened it and walked towards the house. As I was nearing the broad verandah, a man appeared from round the side of the house. He was tall, well-built, and dressed in the usual costume of long, wrinkled boots, plus-four-like *bombachas*, a dirty shirt, and a battered felt hat with a wide brim.

'*Buenos dias*,' he said as he approached me.

'*Buenos dias*. I wish to see Señor Caporal. Is he in the house?' I asked.

He came up to me and swept off his hat.

'I am Caporal,' he said. He took my extended hand, clicked his heels together and bowed very slightly. The gesture was not the-atrical, but automatic. He had a fine brown face, with dark eyes that were full of kindness. He sported a carefully trimmed black moustache under his eagle-beak nose, but his cheeks and chin were covered with a black stubble.

'Do you,' I inquired hopefully, 'speak English?'

'But, of course,' he answered at once, in an impeccable accent that must have been the result of some public school. 'I don't speak it very well, you understand, but I manage to converse fairly well. But please don't stand out here. Come inside and have some coffee, and you can tell me what I may do for you.'

He ushered me through the door which led straight into a small dining-room-cum-sitting-room. The floor was highly pol-ished and gay with locally-woven mats, and the few simple pieces of furniture were similarly polished and glowing. He poked his head through another door and called '*Maria, café para dos, por favor*,' and then turned back to me, smiling.

'This is a great pleasure,' he said sincerely, 'I very seldom get a chance to practise my English. But first, will you excuse me for a brief moment? The coffee will be ready soon . . . here are ciga-rettes . . . please make yourself at home.'

He bowed again slightly, and left the room, I took a cigarette absently and then suddenly, with surprise, noticed that the box that contained them was silver, and with a beautiful and intricate design worked on the lid. Looking about the room I noticed other silver objects, a lovely slender-stemmed vase full of scarlet hibiscus flowers; on the sideboard a pair of beautifully-worked candlesticks, and, between them, a massive fruit bowl that must have weighed a couple of pounds when empty. I began to wonder if perhaps the stories about Caporal were true, for these silver articles were not made in Argentina, and together they were worth a lot of money. He came back in an amazingly short space of time, and I saw that he had washed, shaved and changed into clean *bombachas*, boots and shirt.

'Now I am more fitted to entertain you,' he said smiling, as the Indian maid padded into the room with a tray of coffee, 'what may I do for you?'

I explained about having my own zoo in the Channel Isles, and how I had come to Argentina to collect specimens for it, and he was deeply interested. It turned out that, until quite recently, he had had some wild animals as pets, which he had kept to amuse his children, but as the animals grew bigger and less trustworthy he had sent them all down to the zoo in Buenos Aires. This had been some three days before my arrival in Jujuy, so my feelings can be imagined.

'I had two ostrich,' he said, smiling at my glum expression, 'a fox, an ocelot and a wild pig. I am so sorry I sent them. If I had only known that you were coming . . .'

'Never mind,' I said. 'But if you do get anything during the next ten days, would you send it to Calilegua for me, or send me a message and I'll pick it up?'

'But, of course,' he said, 'with the greatest pleasure.'

He poured me out some more coffee, and we talked about other things. He had the most impeccable manners and the air of a man who had not only been used to money but to a position of authority as well. I began to wonder more and more about him, but he was far too well mannered to talk about himself, and instead tried to steer the conversation into channels which he thought would be of interest to me. Then an idea came to me.

During a slight lull in the talk, I turned to him and said:

'Do please excuse me, but ever since you brought me in I have been admiring your candlesticks. They are quite beautiful. I've never seen anything quite like them before.'

His face lit up delightedly.

'Ah, yes, they are beautiful, aren't they,' he said staring at them. 'They are a little bit of the old régime that I managed to keep . . . those and the other bits of silver you see in here.'

I kept silent, but let myself look faintly puzzled.

'You see,' he went on, 'I am Hungarian. My father was chief minister there before the war. But, after the war, when the Communists came, my father was dead, and I had a wife and three children . . . I did not want them to grow up under such a régime. We escaped, and we only managed to bring a few of my family things with us, most of which we had to sell when we reached Buenos Aires in order to eat. I had some difficulty in getting a job: I had only been trained to be a gentleman.'

He smiled at me shyly, as if ashamed at having bored me with his private reminiscences.

'Still,' he said, offering me a cigarette, 'it is nice to have a few things to remind one of the happy parts of one's life. You would, I think, have liked Hungary in those days. There were plenty of animals there then; what shooting parties we used to have. You like shooting, or do you like the animals too much?'

'No, I don't object to shooting,' I said honestly, 'providing that the animals are not exterminated indiscriminately.'

His eyes glowed eagerly.

'Perhaps,' he said tentatively, 'you would like to see some photographs . . . ?' His voice died away on a faintly interrogative note.

I said I would love to see some photographs, and he went quickly into another room and soon reappeared with a large, beautifully-carved oak chest, which he put on the floor.

He pulled up the lid and tipped out on to the mat a huge heap of photographs, which he shuffled through swiftly. Photograph after photograph he pulled out of the pile, thrusting them excitedly into my hands, wanting to try and communicate some of the happiness they brought back to him. To him they represented hunts he would never forget, and the people that figured in the photographs meant little or nothing to him.

'This is the biggest wild boar we ever shot . . . it was during a drive for the King of Sweden . . . see what a magnificent creature, it was almost a shame to shoot him . . . look at those tusks.'

There the monstrous boar lay on the ground, his lip lifted scornfully over his great tusks, while the King of Sweden stood stiffly, gun in hand, behind him.

'Now, look at this. The best duck drive we had, for the Prince of Siam, five hundred-odd brace of ducks . . . it was a wonderful day, the sky was black with ducks . . . like locusts . . . your feet were frozen and your gun barrels were red-hot but you couldn't stop shooting . . .'

So we looked at photographs for an hour or so, photographs of a parade of animals, royalty and nobility. Then, at last, I pulled out from the pile a large flashlight photograph that showed a massive, panelled dining-room. The chandeliers were like half-grown Christmas-trees, turned upside down, hanging over a long table laden with silver and glass. At the table were seated men and women, beautifully dressed, while at the head of the table sat an elderly man, and on his right a bejewelled and turbaned Indian potentate, and on the other side of the table I recognized my host, immaculate in evening dress. It looked like a scene out of *The Prisoner of Zenda*.

'Oh,' he said carelessly, 'that was a banquet we gave for the Maharajah. Now here, look, see these roedeer, what magnificent heads? Only in Hungary did you get heads like that.'

Presently I heard Helmuth honking his horn in the road outside, and, reluctantly, I rose to go. My host shovelled the photographs back into the chest and closed the lid.

'I am so sorry,' he said contritely, 'I have been boring you with my photographs. If my wife had been here, she would have entertained you more amusingly.'

I protested that I had enjoyed them very much, and, as we went out on to the verandah, there was one question I felt I must ask, whether it was good manners or not.

'Tell me, Señor Caporal,' I said, 'don't you ever miss all that? After having that wonderful life, with money and hunting and influential friends? Don't you find Argentina, to say the least, a little dull?'

He looked at me and laughed.

'Señor Durrell,' he said, 'that which I have been showing you is past, like a dream. It was wonderful while it lasted. But now I have a new life. I am saving a bit of money, so that I may send my children to school in Buenos Aires, and I will have enough left over, I hope, to buy a very small *estancia* for myself and my wife when my children are grown up. What more do I want?'

I pondered on this for a moment, while he watched me, smiling.

'Then you like your job here,' I asked, 'managing this *estancia*?'

'But of course,' he said. 'It is much better than the first job I had when I first came to Argentina.'

'What was that?' I inquired curiously.

'Castrating bulls in Cordoba,' he said chuckling.

I walked down to join Helmuth in the car, feeling very thoughtful. It seemed that I had been privileged to spend the last two hours with a very unusual human being: a truly happy man, and one without bitterness.

By now my collection of creatures had grown to such an extent that it was a whole-time job looking after it. No longer could I go off for three or four days at a time and leave poor Edna to cherish my creatures. Also I was busy building cages for those tame animals which, up until now, had either been at complete liberty, or spent their time tethered, but on leashes. I had originally intended to fly my collection back to Buenos Aires, but the air freight estimate, when it arrived, looked as if it had been worked out by the Astronomer Royal in light-years.

There was nothing for it, I would have to go by train, a two-day and three-night journey that I did not relish, but there was no alternative. Charles arranged the whole thing for me with a speed and efficiency that was typical of him. This in spite of the fact that he had his own work to do, as well as being extremely worried over his wife Joan, who was ill in hospital. So I hammered and sawed in the garden, getting cages ready for my train-journey, and keeping a stern eye on those animals which were still loose and therefore liable to get up to mischief.

The biggest of the still un-caged animals were the coatimundis, Martha and Mathias, who, on collars and chains, were tethered under the trees. I am fond of coatimundis, though they are not everyone's idea of the most charming of animals. But I find something very appealing about their long, rubbery, tip-tilted noses, their pigeon-toed, bear-like walk, and the way they hold their long, ringed tails straight up in the air when they move, like furry exclamation marks. In the wild state they are gregarious, travelling through the forest in quite large parties, uprooting logs and stones, snuffling in every nook and cranny with their vacuum-cleaner-like noses for their prey, which may range from beetles to birds and from fruit to mushrooms. Like most small, gregarious mammals they have quite an extensive vocabulary, and the conversations of a troupe of coatimundis would, I am sure, repay investigation. Mathias would converse with me by the hour in a series of bird-like squeaks and trills; if, when investigating a rotten log or a stone, he thought he was nearing a succulent beetle or slug, the sounds would turn to snuffling grunts, pitched in different keys, and interspersed with a strange champing noise made by chattering his teeth together at great speed. When in a rage he would chitter violently, his whole body shaking as if with ague, and give prolonged, piercing whistling cries that would almost burst your ear drums.

Both the coatimundis had fairly long leashes which were attached to a convenient tree. When they had uprooted and investigated every log and stone within the circle of the leash, they were moved to a fresh tree. Every time this happened, Mathias would spend ten minutes or so marking out his circle of territory with the scent gland at the base of his tail. He would solemnly shuffle his way round in a circle, a look of immense concentration on his face, squatting down at intervals to rub his hindquarters on a convenient rock or stick. Having thus, as it were, hoisted the coatimundi equivalent of the flag, he would relax and settle down to the task of beetle-hunting with a clear conscience. If any of the local dogs were so misguided as to approach his territory they never did it a second time. He would walk slowly towards them, champing his teeth alarmingly, his tail erect, stiff as a poker, and puffed up to twice its normal size. Having got within range he would suddenly dart forward in a curious, rolling run, uttering his piercing, ear-splitting screams. This ghastly noise had the effect of undermining the morale of any but the bravest dog, and, when they had hurriedly retreated, Mathias, quietly chattering and trilling to himself, would wend his way round in a circle, re-marking his entire territory. During all this, Martha would be sitting at the extreme limit of her chain, watching Mathias with adoring eyes, and uttering tiny squeaks of encouragement.

All the other creatures I had acquired were doing splendidly. Juanita, the peccary, grew fatter and more charming each day, and lorded it over the parrots. My precious yellow-naped macaws had given me heart-failure by appearing to go into a decline; I eventually discovered that they were not ill but, for some obscure reason, wanted to sleep inside a box at night, a fact that I discovered quite by accident. As soon as they were supplied with a sleeping box their appetites revived and they started to do well. Among the cats the little Geoffroy's was now quite reconciled to captivity, and played such strenuous games of hide-and-seek with his tabby kitten companion, as well as a game they invented which appeared to be called 'Strangle Your Neighbour', that I began to wonder if I would get them to Buenos Aires alive, let alone Jersey. Luna the puma had tamed down a lot, and even condescended to allow me to scratch her behind the ears, while she rumbled contentedly deep in her throat. The poor half-starved ocelot was now fat and glossy. Having lost the apathy of starvation, she was now very full of herself, and regarded the interior of her cage as sacred, so the process of cleaning her out or feeding her was fraught with danger. Thus are one's kindnesses sometimes repaid.

Among the new creatures which I had added to my collection were two of the most enchanting members of the monkey tribe, a pair of douroucoulis which had been caught in the forest by an Indian hunter. He had been a very good hunter, but unfortunately I had paid him rather too lavishly for the monkeys, and, overcome by the size of the payment, he had retired to the village and stayed drunk ever since, so these were the last specimens I got from him. There is quite an art in paying the right amount for an animal, and by paying too much you can easily lose a good hunter, for between your camp and the forest always lies a series of gin-shops, and hunters are notoriously weak-willed.

Douroucoulis are the only nocturnal monkeys in the world, and from that point of view alone would be remarkable. But when you add to that the fact that they look like a cross between an owl and a clown, that they are the gentlest of monkeys, and that they spend a lot of time clasped in each other's arms exchanging the most human kisses, then douroucoulis become, so far as I am concerned, irresistible. They have the huge eyes, typical of a nocturnal creature, surrounded by a white facial mask edged with black. The shape of the mouth gives you the impression that they are just about to break into a rather sad, slightly pitying smile. Their backs and tails are a pleasant shade of greenish-grey, and they possess great fluffy shirt-fronts that vary from pale yellow to

deep orange, according to age. In the wilds these monkeys, like
the coatimundis, are gregarious, travelling through the trees with
silent leaps in troupes of ten to fifteen animals. The only time they

make any sound is when feeding, and then they converse among themselves with loud, purring grunts which swell their throats up, or a series of bird-like tweets, cat-like mewing, pig-like snufflings and snake-like hissings. The first time I heard them feeding among the dark trees in the forest I identified them as each of these animals in turn, and then became so muddled I was convinced I had found something new to science. I used to dig large red beetles out of the rotting palm-trees for the douroucoulis, insects of which they were inordinately fond. They would watch my approach with the titbits, their eyes wide, their hands held out beseechingly, trembling slightly, uttering faint squeaks of excitement. They would clasp the wriggling beetles in their hands with the awkward grace of a young child accepting a stick of rock, and chew and scrunch their way through them, pausing now and again to utter squeaks of joy. When the last piece had been chewed and swallowed, they would carefully examine their hands, both back and front, to make sure there was none left, and then examine each other for the same reason. Having convinced themselves that no fragment remained, they would clasp each other and kiss passionately for five minutes or so, in what appeared to be an orgy of mutual congratulation.

It was just after I had acquired these delightful monkeys that I had my second encounter with a curious human being, and my introduction to him was due to Luna. He appeared one morning and said that business was to take him to a place a few miles away from Calilegua. In this village he had to visit he had heard rumours of a man who was interested in animals and even kept them as pets.

'All I can find out is that his name is Coco, and that everyone there says he is *loco*, Gerry,' said Luna. 'But you might like to come and see.'

'All right,' I said, nothing loath to leaving my sweat-provoking carpentry for a while, 'but can you wait until I've cleaned and fed the animals?'

'O.K.,' said Luna, and lay patiently on the lawn scratching Juanita's stomach until I had finished my chores.

The village, when we reached it, proved to be a large, straggling one, with a curious dead air about it. Even the houses, constructed as usual with the off-cuts from tree-trunks, had an ill-kempt, dirty look. Everything looked scruffy and depressed. But everyone appeared to know Coco, for when we inquired in the local bar where he lived a forest of hands directed us, and everyone smiled and said, 'Ah, yes, Coco,' as if they were referring to the village idiot. Following directions we found his house easily enough. It would have been very noticeable anyway, for in comparison to the rest of the village, it gleamed like a gem. It had been carefully whitewashed, so that it shone; its front garden was neatly tended and, incredibly, a real gravel path, neatly raked, led up to the house. I decided that if this was the house of the village idiot, then I very much wanted to meet him. In response to our clapping a slight, dark woman appeared, who looked as though she might be Italian. She admitted to being Coco's wife, but said that he was not at home: he worked during the day at the local saw-mill, which we could hear humming in the distance like all the bees in the universe having a conference. Luna explained my mission and the wife's face lighted up.

'Oh,' she said, 'I will send one of the children to fetch him. He would never forgive me if he missed meeting you. Please come round to the back and wait . . . he will come in a few moments.'

The garden at the back of the house was as well tended as the front, and, to my surprise, contained two well-constructed and spacious aviaries. I peered into them hopefully, but they were both empty. We did not have to wait long for Coco's appearance. He appeared from the path leading to the saw-mill at a brisk trot, and arrived, breathing deeply, in front of us and doffed his straw hat. He was a short, well-built man with coal-black, curly hair and (unusual in Argentina) a thick black beard and moustache, carefully trimmed. His eyes were dark, and shone with eagerness as he held out a well-shaped brown hand to Luna and myself.

'Welcome, welcome,' he said, 'you must excuse, please, my English . . . she is not good for I have no chance to practise.'

The fact that he could speak English at all amazed me.

'You have no idea what this means to me,' he said eagerly, wringing my hand, 'to speak with someone who has an interest in Nature . . . if my wife had not called me I would never have forgiven her . . . I could not believe it when my son told me . . . an Englishman to see me, and about animals, too.'

He smiled at me, his face still slightly awe-stricken at this miracle that had happened. One would have thought that I had come to offer him the Presidency of Argentina. I was so overwhelmed at being greeted like a newly-descended angel that I was almost at a loss for words.

'Well,' said Luna, having obviously decided that he had done his job by bringing one lunatic in contact with another, 'I will go and do my work and see you later.'

He drifted off, humming to himself, while Coco seized my

arm gently, as though it were a butterfly's wing that he might damage, and urged me up the steps and into the living-room of his house. Here his wife had produced wonderful lemonade from fresh lemons, heavily sweetened, and we sat at the table and drank this while Coco talked. He spoke quietly, stumbling occasionally in his English and saying a sentence in Spanish when he realized I knew enough of the language to follow. It was an extraordinary experience, like listening to a man who had been dumb for years suddenly recover the power of speech. He had been living for so long in a world of his own, for neither his wife, children nor anyone in the fly-blown village could understand his interests. To him I was the incredible answer to a prayer, a man who had suddenly appeared from nowhere, a man who could understand what he meant when he said that a bird was beautiful, or an animal was interesting, someone, in fact, who could speak this language that had been so long locked up inside him, which no one around him comprehended. All the time he spoke he watched me with an embarrassing expression, a mixture of awe and fear – awe that I should be there at all, and fear that I might suddenly disappear like a mirage.

'It is the birds that I am particularly studying,' he said, 'I know the birds of Argentina are catalogued, but who knows anything about them? Who knows their courtship displays, their type of nests, how many eggs they lay, how many broods they have, if they migrate? Nothing is known of this, and this is the problem. In this field I am trying to help, as well as I can.'

'This is the problem all over the world,' I said, 'we know what creatures exist – or most of them – but we know nothing of their private lives.'

'Would you like to see the place where I work? I call it my study,' he explained deprecatingly, 'it is very small, but all I can afford . . .'

'I would love to see it,' I said.

Eagerly he led me outside to where a sort of miniature wing

had been built on to the side of the house. The door that led into this was heavily padlocked. As he pulled a key from his pocket to open this he smiled at me.

'I let no one in here,' he explained simply, 'they do not understand.'

Up until then I had been greatly impressed with Coco, and with his obvious enthusiasm for animal life. But now, being led into his study, I was more than impressed. I was speechless.

His study was about eight feet long and six feet wide. In one corner was a cabinet which housed, as he showed me, his collection of bird and small mammal skins, and various birds' eggs. Then there was a long, low bench on which he did his skinning, and nearby a rough bookcase containing some fourteen volumes on natural history, some in Spanish some in English. Under the one small window stood an easel, and on it the half-finished water colour of a bird, whose corpse lay nearby on a box.

'Did you do *that*?' I asked incredulously.

'Yes,' he said shyly, 'you see, I could not afford a camera, and this was the only way to record their plumage.'

I gazed at the half-finished picture. It was beautifully done, with a fineness of line and colouring that was amazing. I say amazing because the drawing and painting of birds is one of the most difficult of subjects in the whole natural history field. Here was work that was almost up to the standards of some of the best modern bird painters I had seen. You could see that it was the work of an untrained person, but it was done with meticulous accuracy and love, and the bird glowed on the page. I had the dead specimen in my hand to compare the painting with, and I could see that this painting was far better than a lot I had seen published in bird books.

He pulled out a great folder and showed me his other work. He had some forty paintings of birds, generally in pairs if there was any sexual difference in the plumage, and they were all as good as the first one I had seen.

'But these are terribly good,' I said, 'you must do something with them.'

'Do you think so?' he inquired doubtfully, peering at the paintings.

'I have sent some to the man in charge of the Museum at Cordoba, and he liked them. He said we should have a small book printed when I have enough of them, but this I think is doubtful, for you know how costly a production would be.'

'Well, I know the people in charge of the Museum at Buenos Aires,' I said. 'I will speak to them about you. I don't guarantee anything, but they might be able to help.'

'That would be wonderful,' he said, his eyes shining.

'Tell me,' I said, 'do you like your work here in the saw-mill?'

'Like it?' he repeated incredulously, '*like it*? Señor, it is soul-destroying. But it provides me with enough to live on, and by careful saving I have enough left over to buy paints. Also I am saving to buy a small ciné-camera, for however skilful you are as a painter there are certain things that birds do which can only be captured on film. But these ciné-cameras are very expensive, and I am afraid it will be a long time before I can afford it.'

He talked on for an hour or so, quickly, enthusiastically, telling me what he had accomplished and what he hoped to do. I had to keep reminding myself that this was a man – a peasant, if you prefer the term – who worked in a saw-mill and lived in a house which, though spotless, no so-called 'worker' in England would be seen dead in. To have discovered Coco in the outskirts of Buenos Aires would not have been, perhaps, so incredible, but to find him here in this remote, unlikely spot, was like suddenly coming across a unicorn in the middle of Piccadilly. And, although he explained to me the difficulties of saving enough money to buy paints, and enough to buy his dream ciné-camera, there was never once the slightest suggestion that financial aid might be forthcoming from me. He was simply, with the naïveté of a child, discussing his problems with someone he felt would understand

and appreciate what he was doing. To him I must have represented a millionaire, yet I knew that if I offered him money I would cease to be his friend, and become as the other inhabitants of the village, a person who did not understand. The most I could do was to promise to speak to the Museum in Buenos Aires (for good bird-painters are not two a penny) and to give him my card, and tell him that if there was anything that he wanted from England which he could not obtain in Argentina, to let me know and I would send it to him. When, eventually, Luna reappeared and we simply had to leave, Coco said goodbye wistfully, rather like a child who had been allowed to play with a new toy, and then had it taken away. As we drove off he was standing in the centre of the dusty, rutted street, watching the car and turning my card over and over in his hands, as if it were some sort of talisman.

Unfortunately, on my way down to Buenos Aires I lost Coco's address, and I did not discover the loss until I got back to England. But he had mine, and I felt sure that he would write and ask me to send him a bird book or perhaps some paints, for such things are hard to get in Argentina. But there was no word from him. Then, when Christmas came, I sent him a card, and in it I reiterated my offer to send him anything he needed. I sent the card care of Charles, at Calilegua, who kindly drove out and delivered it to Coco. Then Coco wrote to me, a charming letter, in which he apologized for his bad English, but he thought that, nevertheless, it was improving slightly. He gave me news of his birds and his painting. But there was not a single request in the letter. So, at the risk of offending him, I packed up a parcel of books that I thought would be of the greatest use to him, and shipped them off. And now, when I get disgruntled with my lot, when I get irritated because I can't afford some new animal, or a new book, or a new gadget for my camera, I remember Coco in his tiny study, working hard and enthusiastically with inadequate tools and money, and it has a salutary effect on me. On the way back to

Calilegua Luna asked me what I had thought of Coco, since eve-ryone in the village thought he was *loco*. I said he was, in my opinion, one of the sanest men I had ever met, and certainly one of the most remarkable. I hope that some day I have the privilege of meeting him again.

On the way back to Calilegua we stopped briefly at another village where Luna had heard a rumour that some *bicho* was being kept. To my delight it turned out to be a fully adult male peccary, ridiculously tame, and the perfect mate for Juanita. He was called, by his owner, Juan, and so we purchased him, and put him, gruntling excitedly, into the back of the car and drove back to Calilegua in triumph. Juan, however, was so large, clumsy and eagerly tame that I felt he might, in all innocence, damage Juanita, who was only a quarter his size, and very fragile, so I was forced to cage them separately until Juanita had grown sufficiently. However, they touched noses through the bars, and seemed to be delighted with each other, so I had hope of eventually arranging a successful marriage.

At last the day came when I had to leave Calilegua. I did not want to leave a bit, for everyone had been too kind to me. Joan and Charles, Helmuth and Edna, and the human nightingale Luna, had accepted me, this complete stranger, into their lives, allowed me to disrupt their routine, showered me with kindness and done everything possible to help in my work. But, although I had been a stranger on arrival in Calilegua, such was the kindness shown to me that within a matter of hours I felt I had been there for years. To say that I was sorry to leave these friends was put-ting it mildly.

My journey was, in the first stages, slightly complicated. I had to take the collection on the small railway that ran from Calilegua to the nearest big town. Here everything had to be transhipped on to the Buenos Aires train. Charles, realizing that I was worried about the transhipment side of the business, insisted that Luna travelled with me as far as the main town, and that he, Helmuth and Edna (for Joan was still ill) would drive into the town and meet us there, to sort out any difficulties. I protested at the amount of trouble this would put them to, but they shouted my protests down, and Edna said that if she was not allowed to see me off she would give me no more gin that evening. This frightful threat squashed my protests effectively.

So, on the morning of departure, a tractor dragging a giant flat cart arrived outside Charles's house, and the crates of animals were piled up on it and then driven slowly to the station. Here we stacked them on the platform, and awaited the arrival of the train. I felt distinctly less happy when I examined the railway lines. They were worn right down, obviously never having been replaced for years. In places the weight of the train had pushed both the sleeper and the track so far into the ground that, from certain angles, they seemed to disappear altogether. There was such a riot of weeds and grass growing all over the track anyway, that it was difficult enough to see where the railway began and the undergrowth ended. I estimated that if the train travelled at anything more than five miles an hour on such a track we were going to have the train crash of the century.

'This is nothing,' said Charles proudly, when I protested at the state of the lines, 'this is good compared to some parts of the track.'

'And I thought the plane I came in was dangerous enough,' I said, 'but this is pure suicide. You can't even call them railway *lines*, they're both so bent they look like a couple of drunken snakes.'

'Well, we haven't had an accident yet,' said Charles. And with this cheerful news I had to be content. When the train eventually came into view it was so startling that it drove all thoughts about the state of the track out of my head. The carriages were wooden, and looked like the ones you see in old Western films. But it was the engine that was so remarkable. It was obviously an old one, again straight out of a Wild West film, with a gigantic cowcatcher in front. But someone, obviously dissatisfied with its archaic appearance, had attempted to liven it up a bit, and had streamlined it with sheets of metal, painted in broad orange, yellow and scarlet stripes. It was, to say the least, the gayest engine I had ever seen; it looked as though it had just come straight from a carnival as it swept down towards us at a majestic twenty miles an hour, the overgrown track covering the rails so successfully that the thing looked as though it was coming straight across country. It roared into the station with a scream of brakes, and then proudly let out a huge cloud of pungent black smoke that enveloped us all. Hastily we pushed the animal crates into the guard's van, Luna and I went and got ourselves a wooden seat in the compartment next door, and then, with a great jerk and a shudder, the train was off.

For most of the way the road ran parallel with the railway, only separated by a tangle of grass and shrubs and a low barbed-wire fence. So Charles, Helmuth and Edna drove along parallel with our carriage, shouting insults and abuse at us, shaking their fists and accusing Luna and myself of a rich variety of crimes. The other passengers were at first puzzled, and then, when they realized the joke, they joined in heartily, even suggesting a few choice insults we could shout back. When Helmuth accused Luna of having a voice as sweet as that of a donkey suffering from laryngitis, the orange that Luna hurled out of the train window missed Helmuth's head by only a fraction of an inch. It was childish, but it was fun, and the whole train joined in. At each of the numerous little stations we had to stop at, the idiots in the car would drive on ahead, and be there on the platform to present me with a huge bouquet of wilting flowers, after which I would make a long and impassioned speech in modern Greek out of the train window, to the complete mystification of the passengers who had only just joined the train, and obviously thought that I was some sort of visiting politician. So we enjoyed ourselves hugely until we reached the town where I was to change trains.

Here we piled the collection carefully on the platform, posted a porter in charge to keep people from annoying the animals, and went to have a meal, for there were several hours to wait before the Buenos Aires train got in.

When we reassembled dusk had fallen, and the Buenos Aires train puffed and rumbled its way into the station in an impressive cloud of sparks and steam. But it was just an ordinary engine, and bore not the remotest resemblance to the vivid, lurching dragon that had transported us so nobly from Calilegua. Helmuth, Luna and I carefully stacked the animals into the van that I had hired, and which proved to be far smaller than I anticipated. Charles, meanwhile, had run my sleeping berth to earth, and put my things inside. I was to share it with three other people, but none of them was present, and so I could only hope that they would be interesting. Then, with nothing to do but wait for departure, I squatted on the steps leading down from the carriage, while the others gathered in a group around me. Edna fumbled in her bag, and then held up something that glinted in the dim lights of the station. A bottle of gin.

'A parting present,' she said, grinning at me wickedly, 'I could not bear to think of you travelling all that way without any food.'

'Helmuth,' I said, as Luna went off in search of tonic water and glasses. 'You have a wife in a million.'

'Maybe,' said Helmuth gloomily, 'but she only does this for you, Gerry. She never gives me gin when I go away. She just tells me that I drink too much.'

So, standing on the station, we toasted each other. I had just finished my drink when the guard's whistle squealed, and the train started to move. Still clutching their drinks the others ran alongside to shake my hand, and I nearly fell out of the train kissing Edna goodbye. The train gathered speed, and I saw them in a group under the dim station lights, holding up their glasses in a last toast, before they were lost to view, and I went gloomily to my compartment, carrying the remains of the gin.

The train journey was not quite as bad as I had anticipated, although, naturally, travelling on an Argentine train with forty-odd cages of assorted livestock is no picnic. My chief fear was that during the night (or day) at some station or other, they would shunt my carriage-load of animals into a siding and forget to reattach it. This awful experience had once happened to an animal-collector friend of mine in South America, and by the time he had discovered his loss and raced back to the station in a hired car, nearly all his specimens were dead. So I was determined that, whenever we stopped, night or day, I was going to be out on the platform to make sure my precious cargo was safe. This extraordinary behaviour of leaping out of my bunk in the middle of the night puzzled my sleeping companions considerably. They were three young and charming footballers, returning from Chile where they had been playing. As soon as I explained my actions to them, however, they were full of concern at the amount of sleep I was losing, and insisted on taking turns with me during the night, which they did dutifully during the rest of the trip. To them the whole process must have appeared ludicrous in the extreme, but they treated the matter with great seriousness, and helped me considerably.

Another problem was that I could only get to my animals when the train was in a station, for their van was not connected by the corridor to the rest of the train. Here the sleeping car attendant came into his own. He would warn me ten minutes before we got to a station, and tell me how long we were going to stay there. This gave me time to wend my way down the train until I reached the animal van, and, when the train pulled up, to jump out and minister to their wants.

The three carriages I had to go through to reach the animal van were the third-class parts of the train, and on the wooden benches therein was a solid mass of humanity surrounded by babies, bottles of wine, mothers-in-law, goats, chickens, pigs, baskets of fruit, and other necessities of travel. When this gay, exuberant, garlic-breathing crowd learned the reason for my curious and constant peregrinations to the van at the back, they united in their efforts to help. As soon as the train stopped they would help me out on to the platform, find the nearest water-tap for me, send their children scuttling in all directions to buy me bananas or bread or whatever commodity was needed for the animals, and then, when I had finished my chores, they would hoist me lovingly on board the slow-moving train, and make earnest inquiries as to the puma's health, or how the birds were standing up to the heat, and was it true that I had a parrot that said '*hijo de puta*'? Then they would offer me sweetmeats, sandwiches, glasses of wine or pots of meat, show me their babies, their goats or chickens or pigs, sing songs for me, and generally treat me as one of the family. They were so charming and kind, so friendly, that when we eventually pulled slowly into the huge, echoing station at Buenos Aires, I was almost sorry the trip was over. The animals were piled into a lorry, my hand was wrung by a hundred people, and we roared off to take the creatures, all of whom had survived the journey remarkably well, to join the rest of the collection in the huge shed in the Museum grounds.

That evening, to my horror, I discovered that a good friend of mine was giving a cocktail party to celebrate my return to Buenos Aires. I hate cocktail parties, but could think of no way of refusing this one without causing offence. So, tired though we were, Sophie and I dolled up and we went. The majority of people there I had never met, and did not particularly want to, but there was the sprinkling of old friends to make it worthwhile. I was standing quietly discussing things of mutual interest with a friend of mine when I was approached by a type that I detest. It is the typical Englishman that seems, like some awful weed, to flourish best in foreign climes. This particular one I had met before, and had not liked. Now he loomed over me, wearing, as if to irritate me still further, his old school tie. He had a face empty of expression, like a badly-made death-mask, and the supercilious, drawling voice that is supposed to prove to the world that even if you have no brains you were well brought up.

'I hear,' he said condescendingly, 'that you've just got back from Jujuy.'

'Yes,' I said shortly.

'By train?' he inquired, with a faint look of distaste.

'Yes,' I said.

'What sort of trip down did you have?' he asked.

'Very nice . . . very pleasant,' I said.

'I suppose there was a very ordinary crowd of chaps on the train,' he said commiseratingly. I looked at him, his dough-like face, his empty eyes, and I remembered my train companions: the burly young footballers who had helped me with the night watches; the old man who had recited *Martín Fierro* to me until, in self-defence, I had been forced to eat some garlic too, between the thirteenth and fourteenth stanzas; the dear old fat lady whom I had bumped into and who had fallen backwards into her basket of eggs, and who refused to let me pay for the damage because, as she explained, she had not laughed so much for years. I looked at this vapid representative of my kind, and I could not resist it.

'Yes,' I said sorrowfully. 'They were a very ordinary crowd of chaps. Do you know that only a few of them wore ties, *and not one of them could speak English?*'

Then I left him to get myself another drink. I felt I deserved it.

The Customs of the Country

When you have a large collection of animals to transport from one end of the world to the other you cannot, as a lot of people seem to think, just hoist them aboard the nearest ship and set off with a gay wave of your hand. There is slightly more to it than this. Your first problem is to find a shipping company who will agree to carry animals. Most shipping people, when you mention the words 'animal cargo' to them grow pale, and get vivid mental pictures of the Captain being eviscerated on the bridge by a jaguar, the First Officer being slowly crushed in the coils of some enormous snake, while the passengers are pursued from one end of the ship to the other by a host of repulsive and deadly beasts of various species. Shipping people, on the whole, seem to be under the impression you want to travel on one of their ships for the sole purpose of releasing all the creatures which you have spent six hard months collecting.

Once, however, you have surmounted this psychological hurdle, there are still many problems. There are consultations with the Chief Steward as to how much refrigerator space you can have for your meat, fish and eggs, without starving the passengers in consequence; the Chief Officer and the Bosun have to be consulted on where and how your cages are to be stacked, and how they are to be secured for rough weather, and how many ship's tarpaulins you can borrow. Then you pay a formal call on the Captain and, generally over a gin, you tell him (almost with tears in your eyes) you will be so little trouble aboard that he won't even notice you are there – a statement which neither he nor you believe. But, most important of all, you generally have to have your collection ready for embarkation a good ten days or so before the ship is scheduled to leave, for a number of things may happen in some ports that will put the sailing date forward, or, more irritatingly, backward, and you have to be on the spot to receive your orders. If there is something like a series of dock strikes to delay a ship, you may be sitting round kicking your heels for a month or more, while your animals' appetites appear to increase in direct proportion to your dwindling finances. The end of a trip is, then, the most harrowing, frustrating, tiring and frightening part. When people ask me about the 'dangers' of my trips I am always tempted to say that the 'dangers' of the forest pale into insignificance as compared with the dangers of being stranded in a remote part of the world with a collection of a hundred and fifty animals to feed, and your money running out.

However, we had now, it seemed, surmounted all these obstacles. A ship had been procured, consultations with the people on board had been satisfactory, food for the animals had been ordered, and everything appeared to be running smoothly. It was at this precise juncture that Juanita, the baby peccary, decided to liven up life for us by catching pneumonia.

The animals, as I have said, were now in a huge shed in the Museum grounds, which had no heating. While this did not appear to worry any of the other animals unduly (although it was the beginning of the Argentine winter and getting progressively colder) Juanita decided to be different. Without so much as a pre-liminary cough to warn us, Juanita succumbed. In the morning she was full of beans, and devoured her food avidly; in the evening, when we went to cover the animals for the night, she looked decid-edly queer. She was, for one thing, *leaning* against the side of her box as if for support, her eyes half-closed, her breathing rapid and rattling in her throat. Hastily I opened the door of the cage and called her. She made a tremendous effort, stood upright shakily, tottered out of the cage and collapsed into my arms. It was in the best cinematic tradition, but rather frightening. As I held her I could hear her breath wheezing and bubbling in her tiny chest, and her body lay in my arms limp and cold.

In order to husband our rapidly decreasing money supplies two friends in Buenos Aires had rallied round and allowed Sophie and me to stay in their respective flats, in order to save on hotel bills. So, while Sophie was ensconced in the flat of Blondie Maitland-Harriot, I was occupying a camp-bed in the flat of one David Jones. At the moment when I discovered Juanita's condi-tion David was with me. As I wrapped her up in my coat I did some rapid thinking. The animal had to have warmth, and plenty of it. But I knew we could not provide it in that great tin barn, even if we lit a bonfire like the Great Fire of London. Blondie already had a sick parrot of mine meditatively chewing the wall-paper off the bathroom in her flat, and I felt it was really carrying friendship too far to ask if I could introduce a peccary as well into her beautifully appointed flat. David had now returned at the double from the Land-Rover whence he had gone to get a blanket to wrap the pig in. In one hand he was clasping a half-bottle of brandy.

'This any good?' he inquired, as I swaddled Juanita in the blanket.

'Yes, wonderful. Look, heat a drop of milk on the spirit stove and mix a teaspoonful of brandy with it, will you?'

While David did this, Juanita, almost invisible in her cocoon of blanket and coat, coughed alarmingly. Eventually, the brandy and milk were ready, and I managed to get two spoonfuls down her throat, though it was a hard job, for she was almost unconscious.

'Anything else we can do?' said David hopefully, for, like me, he had grown tremendously fond of the little pig.

'Yes, she's got to have a whacking great shot of penicillin and as much warmth and fresh air as she can get.'

I looked at him hopefully.

'Let's take her back to the flat,' said David, as I had hoped he would. We wasted no more time. The Land-Rover sped through the rain-glistening streets at a dangerous pace, and how we arrived at the flat intact was a miracle. While I hurried upstairs with Juanita, David rushed round to Blondie's flat, for there Sophie had our medicine chest with the penicillin and the hypodermic syringes.

I laid the by now completely unconscious Juanita on David's sofa, and, although the flat was warm with the central heating, I turned on the electric fire as well, and then opened all the windows that would not create draughts. David was back in an incredibly short space of time, and rapidly we boiled the hypodermic and then I gave Juanita the biggest dose of penicillin I dared. It was, almost, kill or cure, for I had never used penicillin on a peccary before, and for all I knew they might be allergic to it. Then, for an hour, we sat and watched her. At the end of that time I persuaded myself that her breathing was a little easier, but she was still unconscious and I knew she was a very long way from recovery.

'Look,' said David, when I had listened to Juanita's chest for the fourteen-hundredth time, 'are we doing any good, just sitting here looking at her?'

'No,' I said reluctantly, 'I don't think we'll really see any change for about three or four hours, if then. She's right out at the moment, but I think the brandy has a certain amount to do with that.'

'Well,' said David practically, 'let's go and get something to eat at Olly's. I don't know about you, but I'm hungry. We needn't be more than three-quarters of an hour.'

'O.K.,' I said reluctantly, 'I suppose you're right.'

So, having made sure that Juanita was comfortable and that the electric fire could not set fire to her blankets, we drove down to Olly's Music Bar in 25 de Mayo, which is a street that runs along what used to be the old waterfront of Buenos Aires. It is a street lined with tiny clubs, some of which have the most delightful names like 'My Desire', 'The Blue Moon Hall of Beauties', and, perhaps slightly more mysteriously, 'Joe's Terrific Display'.

It was not the sort of street a respectable man would be seen in, but I had long ceased to worry about respectability. With my various friends we had visited most of these tiny, dark, smoky bars, and drunk drinks of minute size and colossal price, and watched the female 'hostesses' at their age-old work. But, of all the bars, the one we liked best was Olly's Music Bar, and we always made this our port of call. There were many reasons for liking Olly's. Firstly, was the walnut-wrinkled Olly himself and his lovely wife. Secondly, Olly not only gave you fair measure in your glass, but frequently stood you a drink himself. Thirdly, his bar was well-lit, so that you could actually see your companions; in the other bars you would have had to be a bat or an owl to observe clearly. Fourthly, his hostesses were not allowed to irritate you by constantly suggesting you bought them drinks, and fifthly, there was a brother and sister with a guitar who sang and played delightfully. Lastly, and perhaps most importantly, I have seen the hostesses at Olly's, when their night's work was done, kiss Olly and his wife goodnight as tenderly as if they had been the girls' parents.

So David and I made our way down the stairs into Olly's, and were greeted with delight by Olly and his wife. The reason for our depression being explained the whole bar was full of commiseration; Olly stood us both a large vodka, and the hostesses gathered round to tell us how they were sure Juanita would get well, and generally tried to cheer us up. But, as we stood there eating hot sausages and sandwiches and consuming vodka, not even the gay *carnavalitos* the brother and sister played and sang specially for us could cure my depression. I felt sure that Juanita was going to die, and I had grown absurdly fond of the little creature. Eventually, when we had eaten and drunk, we said goodbye and climbed the steps that led to the road.

'Come tomorrow and tell us how the animal is,' called Olly.

'*Si, si,*' said the hostesses, like a Greek chorus, 'come tomorrow and tell us how the *pobrecita* is.'

By the time we had got back to David's flat I was convinced that we should find Juanita dead. When we went into the living-room I gazed at the pile of blankets on the sofa, and had to force myself to go and look. I lifted one corner of the blanket gently and a twinkling dark eye gazed up at me lovingly, while a pink plunger-shaped nose wiffled, and a faint, very faint, grunt of pleasure came from the invalid.

'Good God, she's better,' said David incredulously.

'A bit,' I said cautiously. 'She's not out of danger yet, but I think there's a bit of hope.'

As if to second this Juanita gave another grunt.

In order to make sure that Juanita did not kick off her blanket during the night and make her condition worse I took her to bed with me on the sofa. She lay very quietly across my chest and slept deeply. Though her breathing was still wheezy it had lost that awful rasping sound which you could hear with each breath she took to begin with. I was awoken the following morning by a cold, rubbery nose being pushed into my eye, and hearing Juanita's wheezy grunts of greeting. I unwrapped her and saw she was a different animal. Her eyes were bright, her temperature was normal, her breathing was still wheezy, but much more even, and, best of all, she even stood up for a brief, wobbly moment. From then she never looked back. She got better by leaps and bounds, but the better she felt the worse patient she made. As soon as she could walk without falling over every two steps, she insisted on spending the day trotting about the room, and was most indignant because I made her wear a small blanket, safety-pinned under her chin, like a cloak. She ate like a horse, and we showered delicacies on her. But it was during the nights that I found her particularly trying. She thought this business of sleeping with me a terrific idea, and, flattering though this was, I did not agree. We seemed to have different ideas about the purposes for which one went to bed. I went in order to sleep, while Juanita thought it was the best time of the day for a glorious romp. A baby peccary's tusks and hooves are extremely sharp, and their noses are hard, rubbery and moist, and to have all these three weapons applied to one's anatomy when one is trying to drift off into a peaceful sleep is trying, to put it mildly. Sometimes she would do a sort of porcine tango with her sharp hooves along my stomach and chest, and at other times she would simply chase her tail round and round, until I began to feel like the unfortunate victim in *The Pit and the Pendulum*. She would occasionally break off her little dance in order to come and stick her wet nose into my eye, to see how I was enjoying it. At other times she would become obsessed with the idea that I had, concealed about my person somewhere, a rare delicacy. It may have been truffles for all I know, but whatever it was she could make a thorough search with nose, tusks and hooves, grunting shrilly and peevishly when she couldn't find anything. Round about three a.m. she would sink into a deep, untroubled sleep. Then, at five-thirty, she would take a quick gallop up and down my body to make sure I woke up in good shape. This lasted for four soul-searing nights, until I felt she was sufficiently recovered, and then I banished her to a box at night, to her intense and vocal indignation.

I had only just pulled Juanita round in time, for no sooner was she better than we got a message to say that the ship was ready to leave. I would have hated to have undertaken a voyage with Juanita as sick as she had been, for I am sure she would have died.

So, on the appointed day, our two lorry-loads of equipment

and animal-cages rolled down to the dock, followed by the Land-Rover, and then began the prolonged and exhausting business of hoisting the animals on board, and arranging the cages in their places on the hatch. This is always a nerve-racking time, for as the great nets, piled high with cages, soar into the air, you are always convinced that a rope is going to break and deposit your precious animals either into the sea or else in a mangled heap on the dock-side. But, by evening, the last cage was safely aboard, and the last piece of equipment stowed away in the hold, and we could relax.

All our friends were there to see us off, and, if in one or two people's eyes was a semi-repressed expression of relief, who was to blame them, for I had made martyrs of them all in one way or another. However, we were all exhausted but relaxed, ploughing our way through a series of bottles I had had the foresight to order in my cabin. Everything was on board, everything was safe, and now all we had to do was to have a farewell drink, for in an hour the ship was sailing. Just as I was replenishing everyone's glass for the fifth toast, a little man in Customs uniform appeared in the cabin doorway, rustling a sheaf of papers. I gazed at him fondly, without any premonition of danger.

'Señor Durrell?' he asked politely.

'Señor Garcia?' I inquired.

'*Si*,' he said, flushing with pleasure that I should know his name, 'I am Señor Garcia of the Aduana.'

It was Marie who scented danger.

'Is anything wrong?' she asked.

'*Si, si, señorita*, the señor's papers are all in order, but they have not been signed by a *despachante*.'

'What on earth's a *despachante*?' I asked.

'It is a sort of man,' said Marie worriedly, and turned back to the little Customs man. 'But is this essential, señor?'

'*Si, señorita*,' he said gravely, 'without the *despachante's* signature we cannot let the animals be taken. They will have to be unloaded.'

I felt as though someone had removed my entire stomach in one piece, for we had about three-quarters of an hour.

'But is there no *despachante* here who will sign it?' asked Marie.

'Señorita, it is late, they have all gone home,' said Señor Garcia.

This is, of course, the sort of situation which takes about twenty years off your life. I could imagine the shipping company's reaction if we now went to them and told them that, instead of gaily casting off for England in an hour's time, they would be delayed five hours or so while they unloaded all my animals from the hatch, and, what was worse, all my equipment and the Land-Rover which were deep in the bowels of the ship. But by now my friends, unfortunate creatures, were used to crises like this, and they immediately burst into activity. Mercedes, Josefina, Rafael and David went to argue with the Chief of Customs on duty, while Willie Anderson, another friend of ours, went off with Marie to the private home of a *despachante* he knew. This was on the outskirts of Buenos Aires, so they would have to drive like the devil to get back in time. The happy farewell party burst like a bomb and our friends all fled in different directions. Sophie and I could only wait and hope, while I mentally rehearsed how I would phrase the news to the Captain, without being seriously maimed, if we had to unload everything.

Presently the party who had been arguing with the Chief of Customs returned despondently.

'No use,' said David, 'he's adamant. No signature, no departure.'

'He is very much what you call a stupid buggler,' said Josefina, and then, struck by a thought, 'Gerry, tell me what does this word buggler mean? I look up in dictionary and all I find is a man who plays a buggle. This is not insulting, no?'

But I was in no condition to help Josefina out with her English translations. We had twenty minutes to go. At that moment we heard a car screech to a halt on the docks outside. We piled out on to the deck, and there, coming up the gangway, smiling triumphantly, were Marie and Willie, waving the necessary documents, all beautifully signed by what must be the finest, noblest *despachante* in the business. So, with ten minutes to go we all had a drink. I even gave Señor Garcia one.

Then the steward poked his head in to say that we would be casting off in a moment, and we trooped on to the deck. We said our goodbyes, and our tribe of friends made their way down on to the quay. Ropes were cast off, and slowly the gap between the ship and the dock widened, so that we could see the shuddering reflection of the quay lights in the dark waters. Presently the ship gained speed, and soon our friends were lost to sight, and all we could see was the great heap of multicoloured lights that was Buenos Aires.

As we turned away from the rail and made our way to our cabins, I remembered Darwin's words, written a century before. When speaking of the travelling naturalist he said: '*he will discover how many truly kind-hearted people there are with whom he had never before had, or ever again will have, any further communication, who yet are ready to offer him the most disinterested assistance.*'

Stop Press

For those that are interested here is an up-to-date account of the creatures we brought back. Claudius the tapir, whom I could once lift up in my arms – at the risk of a rupture – is now the size of a pony, and eagerly awaiting a bride when we can afford one.

Mathias and Martha, the coatimundis, have settled down to domestic bliss and have produced two sets of children. Martha, at the time of writing, is again in an interesting condition.

Juan and Juanita, the peccaries, also had two sets of babies, and are expecting a third.

Luna, the puma, the ocelot and the Geoffroy's cat are all flourishing, getting fatter with each passing day.

Blanco, the Tucuman Amazon, still says '*Hijo de puta*', but very softly now.

All the other birds, beasts and reptiles are equally well, and many showing signs of wanting to breed.

Which leaves me with only one thing to say and thus, I hope, stop people writing to ask me: my zoo is a private one, but it is open to the public every day of the year except Christmas Day.

So come and see us.

Acknowledgements

As always after an expedition there are those people to whom one's gratitude is so immense that there is no way of adequately thanking them. All I can do is reiterate once more how much I appreciated their help and encouragement.

Buenos Aires

The entire de Sota family; the entire Rodrigues family; our dear friend Bebita Ferreyra; Lassie Greenslet; David Jones; Josefina Pueyrredon; Dicky de Sola; Brian Dean; Bill Partridge; and Willie Anderson.

All these people assisted us in countless ways, giving advice and helping us clear our equipment through the Customs; entertaining us lavishly and acting as drivers, translators, guides, carpenters and cooks, on our behalf.

People whose patience we tried, and whose houses and places of work we infested with our animals are: Blondie Maitland-Harriot; Mrs Dorothy Krotow; Dr Mario Teruggi. To them all we – and our animals – are most grateful.

To Dr Carlos Godoy, my special thanks, as he was so efficient and helpful over our collecting permits and in furnishing us with letters of introduction to many throughout the Argentine.

Dr Caberra was extremely helpful in giving us information regarding the Argentine fauna.

Mr Salmon of Bovril, Ltd was most kind and helpful. Mr Blackburn of Chadwick Weir arranged for the transportation of the entire collection and equipment from the Argentine, a massive undertaking.

Puerto Deseado

To Señor Huichi for his help we simply cannot express our gratitude enough. Captain Giri was instrumental in introducing us to
Señor Huichi and for helping us find the penguin colonies. For
both of these things we were most grateful. Mr Bateman, the
British Vice-Consul, and his wife assisted us in every way possible, as did Mr and Mrs Roberts, the local Postmaster and his wife.
All these people did their utmost to make our stay in Deseado a
pleasant one.

Puerto Madryn

The manager of the Hotel Playa not only provided us with
accommodation but lent us money, sent telegrams for us and
helped us in every other way he could.

Jujuy

Charles and Joan Lett; Edna and Helmuth Vorbach; Luna, a very
good friend, and everyone at Calilegua accepted me into their
midst and did everything to help me build up my collection of
animals, make my film and arrange everything for my comfort
and salvation. Without them all I would have been lost.

Mendoza

Dr Menoprio, who was so kind to us in many ways.

Britain

Mr Peter Newborne of C.A.P., who was his usual helpful self and did all he could to assist us in the complicated matters of Customs facilities, etc. Dr Don Alberto Candiotti the former Argentine Ambassador in London, who gave the whole expedition his official blessing and encouraged us in every way. Mr Lawton Johnson of Bovril arranged for us to visit the various Bovril estates in the Argentine, which, unfortunately, we were unable to do; Mr Flack and Mr Aggett, of Blue Star Line, arranged passages for us all. The South American Saint Line kindly agreed to transport my entire collection and all my equipment from Buenos Aires to England, and in this connection I would like to thank the Captain and crew of the M.V. *St John* for enabling the return voyage to go smoothly, which was due entirely to their help and kindness.

The following British manufacturers supplied us with various equipment, without which the expedition would have been a complete failure. The Rover Company supplied us with our Land-Rover, in which we were able to travel all over the Argentine, and Mr Baldwin and Mr Bradley of the Company's Sales and Publicity Dept were extremely kind and helpful in enabling us to have this vehicle.

The Directors of William Smith (Poplar) Ltd, The British Nylon Spinners Ltd, and Greengate & Irwell Rubber Co. Ltd continue to earn our deep gratitude for the wonderful tarpaulins and animal shelters that they gave us on a previous trip. These articles, which are in constant use, have proved absolutely invaluable.

Finally, may we thank all those both here and in the Argentine who have helped us in many small ways, but without whose help the expedition could not have been successful.

GERALD DURRELL

AFTERWORD

A message from the Durrell Wildlife Conservation Trust

The end of this book isn't the end of Gerald Durrell's story. The various experiences you have just read about gave impetus and inspiration to his lifetime crusade to preserve the rich diversity of animal life on this planet.

Although he died in 1995, the words of Gerald Durrell in this and his other books will continue to inspire people everywhere with love and respect for what he called 'this magical world'. His work goes on through the untiring efforts of the Durrell Wildlife Conservation Trust.

Over the years many readers of Gerald Durrell's books have been so motivated by his experiences and vision that they have wanted to continue the story themselves by supporting the work of his Trust. We hope that you will feel the same way today because through his books and life, Gerald Durrell set us all a challenge. 'Animals are the great voteless and voiceless majority', he wrote, 'who can only survive with our help'.

Please don't let your interest in conservation end when you turn this page. Write to us now and we'll tell you how you can be part of our crusade to save animals from extinction.

For further information, or to send a donation, write to:

Durrell Wildlife Conservation Trust
Les Augrès Manor
Jersey, Channel Islands, JE3 5BP
UK
www.durrell.org